Critical praise for *Awak*

Going beyond a new light bulb is crucial if we're to make an impact on climate change—Carla Wise offers some sage insights in this new volume.

—Bill McKibben, author of *Eaarth*

This is the best, most even-handed and honestly researched general guide to the current state of climate change that you are likely to find. *Awake on Earth* is a highly readable balance of science and personal experience, raising tantalizing possibilities for human transformation.

—Kathleen Dean Moore, author of *Great Tide Rising* and co-editor of *Moral Ground*

The qualities that make Carla Wise's book important to read are clarity, concision, and a scientifically informed and humane sense of judgment and balance, such that all the issues in this huge topic ... are laid out in their proper proportion and sequence. Reading this book will fill in any blanks the reader may have, and altogether it offers the sort of cognitive mapping that gives the reader orientation for action, and also hope.

—Kim Stanley Robinson, author of the *Mars trilogy* and *Green Earth*

(continued on next page)

We have plenty of scientific studies on the climate catastrophe, but no guidance on how we, as human beings, gather the courage and collective spirit to rise to this moment—until now. Carla Wise writes with the voice of a close friend taking us on a journey through the stark reality to explain the choices we face. A beautifully written and honest appraisal, *Awake on Earth* helps us find our common humanity and face up to the profound moral obligation that rests on our shoulders simply from being alive at this crucial time on Earth. It is a must-read for anyone who loves a child.

—Mary Christina Wood, author of *Nature's Trust: Environmental Law for a New Ecological Age*

This highly important book blends succinct descriptions of the major ecological and human consequences of climate change with the author's compelling and poignant reflections on her own coming to terms with the data.

In it, Carla Wise beautifully spells out the many existential and practical dilemmas of climate change—how is it possible that humans have achieved such prosperity at the same time as things are so dire; what does the reality of climate change mean for how privileged people might struggle with and continue to find meaning in daily life; how do we live well in an age of climate change.

Awake on Earth is a real gem and I absolutely recommend it. I thoroughly enjoyed reading it and was sad when I came to the end.

Well written and inspiring. Simply excellent.

—Kari Norgaard, author of *Living in Denial: Climate Change, Emotions, and Everyday Life*

Awake on Earth

Awake on Earth

Facing Climate Change with Sanity and Grace

Carla A. Wise, Ph.D.

Wise on Earth Press
Corvallis, Oregon

Grateful acknowledgment is made to Counterpoint Press for permission to reprint "The Peace of Wild Things" from *New Collected Poems* by Wendell Berry (Counterpoint Press, copyright © 2012 by Wendell Berry).

Editor: Lorraine Anderson
Front cover photo: Mark Van Steeter
Cover and interior design: Meadowlark Publishing Services

For information:
Wise on Earth Press
3710 NW Wisteria Way
Corvallis, OR 97330

ISBN 978-0-692-68696-6
Printed in the United States of America
First printing: June 2016

For my dad, who taught me so much
through his quiet example.

Contents

Preface ix
Introduction xiii
Starting Premises xix

Part 1: The Signs of a New World 1
1 Extreme Weather. 3
2 Life on Land 11
3 Oceans. 19
4 Forests and Trees. 31
5 Agriculture 41
6 Paradoxes 51

Part 2: The Question of How to Respond 59
7 Blame and Moving On. 63
8 Guilty Pleasures 73
9 Grappling with Despair 79
10 Seeking Solace 85
11 Looking for Solutions 95
12 Spotting Hopeful Trends 107
13 Preparing 121
14 Finding My Way 127

Take Action 141
Acknowledgments 143
Notes 145
Index 175
About the Author 187

Preface

As a biologist, environmental writer, and mother, I'd been concerned about climate change for many years, but around 2006 I became obsessed. Studying and worrying about climate change became a big part of my life. It turns out this isn't how most people choose to spend their free time, and it's not a great way to make friends, either. It did, however, lead me to write *Awake on Earth*. So I've been wondering what can I say on this first page to induce you to read an entire book—even a short one—about my journey coming to terms with and finding a way to respond to climate change.

There are a lot of other climate change books out there. Yet in all my reading, I never found a compelling, understandable, brief yet accurate explanation of human-caused climate change for nonscientists. More important, I couldn't find a thoughtful exploration of how to feel and what to do about it. So I wrote this book, my attempt to do those things, and it changed my life. You should read this book because I believe it will change yours, too.

At this moment, we are on a path toward catastrophically large and rapid climate change that threatens to disrupt all of Earth's ecosystems and all human civilization. The odds are against us getting off this path. One reason (among many) we have failed to grasp this is that anyone who has spoken plainly about the facts has been dismissed, ignored, or labeled an alarmist. It's so difficult to believe what's going on that we try to find reasons to reject all warnings: the science is uncertain, the threat is exaggerated, the message is politically motivated, there is nothing we can do.

No one I know *wants* the threat of human-caused climate change to be as bad (or as imminent) as it is. I hope you will keep reading because I have carefully examined the available evidence and done my very best to tell the honest truth about what we know and what is uncertain. There are many unknowns about the *details* of what is to come. We don't know exactly what the average temperature will be on Earth next year, or in a decade. We don't know how quickly sea levels will rise or exactly how warming will affect all the factors that influence tornadoes. We don't know when certain thresholds might be reached, such as large-scale thawing of permafrost or unstoppable melting of ice sheets, that could lead to irreversible, abrupt, or catastrophic climate change. We do know the broad outlines of what is happening and why, and some details of how warming is likely to play out. What we do know is compelling, amazing, and pretty scary.

But the main reason I hope you will keep reading is that this book is not just about examining the facts but also about responding to them in a constructive, sane, ethical, life-affirming way. We are alive at possibly the most critical—and interesting—moment in human history. We are the first humans to experience the planetwide climatic consequences of our species' actions, and the last humans with the chance to limit those changes to adaptable levels. Awakening fully to what lies before us — the era of human-caused climate change — is the only good option available. Ultimately, I've discovered that to live with optimism and integrity in these times requires both facing that everything we depend on and value is in imminent danger *and* working to improve the chances for a livable future. I'm not suggesting I know the best way for you to do these two things, but I can tell you how I have done them and what this has given me.

If you're ready to understand this era of human-induced climate change, or feel you don't yet know enough about it, or are seeking a way to negotiate these times with sanity and perhaps even some grace, this book is for you. Writing it has helped me find my way, and I hope reading it will help you find yours.

Introduction

I'm old enough to remember a time before climate change was considered a political issue, a time before Republicans running for Congress were afraid to acknowledge they believed humans were influencing the climate. Therefore, I'm old enough to understand that things might have gone very differently. I was in graduate school when the first President Bush took office. Congress had just been briefed about global warming by NASA's chief climate scientist, and the new EPA administrator, William Reilly (a Republican), wanted to do something about it. I remember hearing Reilly give a great speech about his plans and feeling hopeful. If George H. W. Bush had championed climate legislation, or if a few years later Bill Clinton had gotten Congress to ratify the Kyoto Protocol, or if fossil fuel companies had not been so effective in sowing doubt about the scientific evidence, we might be following a different trajectory. Instead, climate change became a battleground between our two major political parties throughout the 1990s.

Throughout the first decade of the twenty-first century, I continued to hope that we would awaken to the nature

> **Global warming** and **climate change** mean essentially the same thing: the recent increase in the atmospheric and oceanic temperatures on planet Earth due to higher concentrations of carbon dioxide, hydrocarbons, and other pollutants in the atmosphere causing an enhanced greenhouse effect.

of the danger before us. Instead, climate change became even more politically charged, and here we are, at this very odd moment in human history. We are still caught in a standoff over how we will respond to an ever-growing threat to our life-support systems. The 2015 Paris climate accord suggests the global community is beginning to agree on what needs to happen, but only the actions of governments and people in the coming years will show if Paris signaled a real turning point. Climate change is now affecting the world's oceans, landscapes, and weather patterns; it is contributing to rising sea levels, species extinctions, crop failures, heat waves, droughts, floods, and wildfires. The best available science tells us we may now be too late to avoid truly civilization-threatening levels of climate change. These are the simple facts. *So where do we possibly go from here?*

I remember the day in early June 2010 when I first asked myself this question and started writing this book. That was the day it hit me that I had experienced a fundamental shift in my view of the possibilities before us. It was the day I realized that I believed that changing course in time to avoid catastrophic levels of climate change was highly unlikely, maybe even impossible.

The information that led to my realization is all around us today in dire scientific, military, and strategic reports, and my belief is reinforced by repeated national and international failures and delays in regulating greenhouse gas emissions, along with increasing climate-related disasters all over the world. The evidence was less glaring in 2010, but it was there if you looked. And I had many reasons to look carefully a little sooner than some. I have been drawn to the natural world my whole life. I grew up running around in the outdoors, raising chickens, riding horses, and hiking; my favorite class in high school was biology, and I studied ecology in college. I have a master's degree in natural resource policy and a Ph.D. in biology. I've worked on an organic farm and in environmental education, natural resource policy analysis, environmental consulting, plant conservation, and environmental writing.

Throughout my varied work life, I've always wanted to contribute to humans coming into better harmony with the living systems that sustain us. On that summer day in 2010, it dawned on me that it was probably too late for my work to make any difference: all the evidence I was seeing pointed toward the likelihood of the climate destabilizing beyond repair. All the topics I wrote about—agriculture, endangered species, forest fire management, endocrine-disrupting chemicals, the local foods movement—still mattered to me. But the premise behind my writing, that I could illuminate and educate, that this could help bring about change, was slipping away. Without a path to stabilizing the climate, I realized no other issues I cared most about could be solved either.

Of course, it wasn't just about my work. All the places I love, places of inspiration and renewal, are changing. These changes threaten the living fabric of these natural wonders. I've had the incredible good fortune of visiting the Amazon jungle, exploring the Galapagos Islands, and floating some of the most magnificent desert rivers in the American Southwest. I've hiked and camped in the Sierra Nevada, the Rocky Mountains, and the Oregon Cascades.

Facing what climate change means for the places I love and the creatures that dwell there was deeply painful.

Finally, there is my daughter, Lia, who was born in the summer of 1999. Motherhood changed me, turning me into someone more selfless and fearful. I had never been afraid of death before or thought much about the inborn spirit in each of us. I became a fiercely protective mom in a way that was almost embarrassing. I felt a love for Lia that was indescribably complete but also painful in its intensity. The future became personal for me. What potentially catastrophic climate change would mean for my daughter was something I could hardly bear to imagine when I started writing this book. I had to find a way.

The journey has sometimes been rocky. What I felt in the beginning was mostly despair, mixed with amazement that we humans could actually be in this fix and largely unaware of it. Yet because of who I am and what I love, and because I believe that at this moment on Earth nothing else matters as much, I found that I needed to understand the era of climate change and find a way to respond. What I provide here is a record of my exploration that is grounded in science, holistic, and personal. In these pages I'm striving to understand what we are facing, why, and how we might respond with more honesty, courage, and even grace.

The book begins by briefly explaining the consensus among essentially all the world's climate scientists about how climate change is proceeding and why. Then the chapters in Part 1 describe how the changing climate is affecting weather extremes, land-based life forms, the oceans, forests, and agriculture. This section of the book is necessarily a bit science-y, but I've done my best to hit just the highlights and reduce the jargon to a minimum.

Part 1 ends with an examination of some of the paradoxes of living in these times.

In Part 2 I search for ways to talk about, think about, and feel the implications of our destabilizing climate. I explore ways to get prepared, take action, and find solace. As I dug into this project, I began to understand that giving up on solving the climate crisis has some rewards. I learned that there are things to do and choices to be made about how to live well in our times.

For me, figuring out how to respond requires facing an out-of-control future and letting go of the hope of acting in time to save the entire world. It has to do with finding ways to take greater responsibility for the survival and health of ourselves, our families, and our communities. It has to do with tapping into a deep sense of the resiliency of the planet and its life, and even the vastness of the universe, whatever happens to us. Finally, it has to do with making the choice to *really do something* and figuring out what that something is. I'll give you a hint: changing your light bulbs and recycling is not what I'm talking about. These extraordinary times call out to us for more.

Starting Premises

Before becoming an environmental writer, I earned a Ph.D. in biology and for more than a decade studied the ecology and genetics of rare plants. My time as a biologist taught me how scientific knowledge is built and how a scientific consensus is formed.

Here's how it goes: starting in graduate school, as you develop research proposals, apply for grants, and submit papers for publication, you must defend each idea, method, and conclusion. Your work is sent out for review to experts in your field who search for flaws in it. Your grant applications and manuscripts frequently get rejected or are returned with a list of criticisms you are required to address for reconsideration. When you publish your findings or give a talk, your research is sometimes criticized or your results disputed. These processes vary in intensity based on the specific field and the personalities involved, but from my observations they happen across disciplines. This is just how science works.

I loved the questions I got to ask as a biologist, but for me, the system took much of the pleasure out of doing science. I lacked the competitive drive and the ego to thrive;

I was too sensitive and took the criticism too personally. However, I learned a vital lesson: if a scientific consensus exists about something, lots of very smart, hardworking, competitive, detail-oriented people have already tried to disprove it, and they have failed. When scientists in an entire field agree on something, it is *almost certain* to be true. This is why it astonishes me to find so little public understanding and acceptance of what climate scientists have learned since the 1980s. What follows is a brief summary of the scientific consensus on climate change and what is known and what is uncertain about the future trajectory of Earth's changing climate.

The science is settled

Greenhouse gases—which include carbon dioxide, methane, nitrous oxide, and water vapor—act like a blanket, holding heat in Earth's lower atmosphere. This phenomenon, called the greenhouse effect, keeps Earth's surface warm enough for life to thrive. However, higher concentrations of greenhouse gases increase the amount of heat held in. Climate scientists have determined that Earth's climate is warming and that human activities that release greenhouse gases into the atmosphere are the primary cause of the observed warming.[1]

This scientific consensus has been built using a variety of approaches over many decades. Scientists have been recording Earth's land and ocean temperatures at multiple locations starting around 1880 and increasing in number and quality ever since. NASA's Goddard Institute for Space Studies maintains a graph of Earth's global mean land-ocean temperature index over time, updated monthly and available online. The graph plotting temperatures from 1880 to 2015 is shown here as Figure 1. This graph

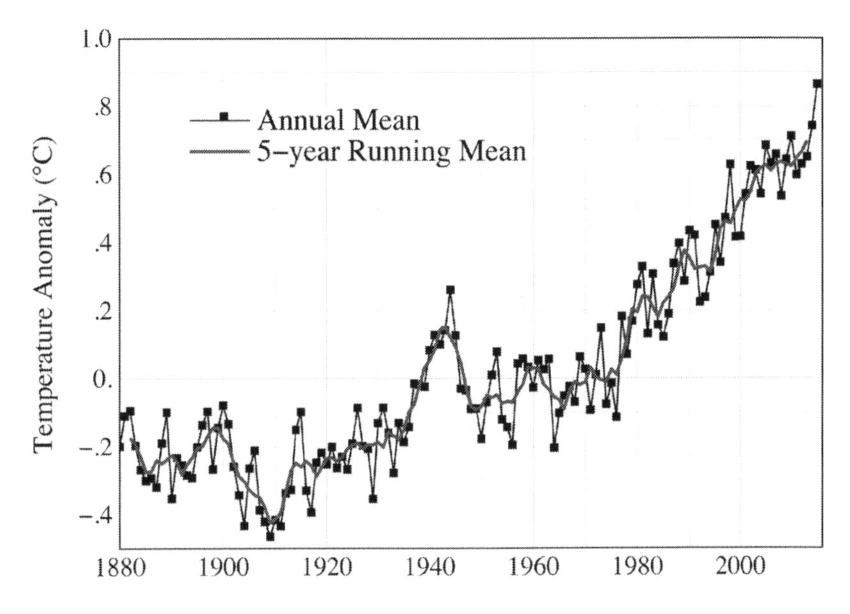

Figure 1. Earth's global mean land-ocean temperature index, from NASA's Goddard Institute for Space Studies. Source: data.giss.nasa.gov/gistemp/graphs_v3/.

uses 1951 to 1980 as the 0-degree baseline and shows the annual mean temperature in squares and the five-year mean as a smoothed line. Analyses of measurements between 1880 and 2012 showed that Earth's average surface temperature rose by approximately 1.5 degrees F (0.85 degrees C)[2] during that period and that the warming trend had been accelerating since 1980.[3] Including temperatures through 2015, average temperatures have risen 1.8 degrees F (1 degree C) since 1880.[4]

Scientists have also been measuring concentrations of greenhouse gases since the 1950s and have used multiple methods to estimate greenhouse gas concentrations before the 1950s. Human activities such as burning fossil fuels and cutting down forests have increased the concentration of carbon dioxide in the atmosphere by 41 percent since the start of the Industrial Revolution and have also increased the concentrations of methane (now at two and a half times preindustrial levels) and nitrous oxide (15 percent above preindustrial levels).[5] Multiple lines of evidence, including numerous studies and climate models, indicate that increasing concentrations of greenhouse gases—not natural phenomena—are the primary cause of recent climate change.

This book is not about climate science—many such books have already been written[6]—but the basic message is pretty simple: human-caused climate change is happening, and this has been proven beyond any reasonable doubt. If you've ever wondered about the objections of climate change deniers, know this: every one of these objections, whether completely spurious or sincere, has been carefully examined and tested by climate scientists. All of the alternative hypotheses have been found to be false. Temperature records are in fact sound and have been refined and reconfirmed. Natural phenomena—El Niño, volcanic activity, variations in solar output—have been examined and cannot explain rising temperatures over the past century. Only rising greenhouse gas levels can.[7] There was no warming hiatus in the 2000s, only variation in the apparent pace of warming, possibly related to natural variability, ocean heat storage, or El Niño events.[8]

Climate change is altering our world

Scientists also agree that climate change is causing a variety of increasingly dramatic impacts worldwide. Heat waves and record high temperatures have increased, while record lows have decreased, and average snowfall has declined in many cold regions. As air warms, it holds more energy and more moisture, which is leading to worsening storms and increasing climate instability—meaning more weather disasters and higher flood risks in many regions.[9]

As the climate has warmed, impacts on living things have been documented throughout the world. Species everywhere have shifted their ranges in response to changing temperatures,[10] and seasonal activities such as spring migration, blooming, and breeding have been shifting earlier since the 1970s.[11] Worldwide, species extinction rates are already estimated to be a hundred to a thousand times above baseline rates due to human activities, but climate change is expected to help push extinction rates much higher in the coming decades.[12]

Rapid warming in the Arctic has led to many dramatic changes there: Arctic sea ice cover has declined approximately 13 percent per decade since measurements began in 1979.[13] Many of the world's glaciers and ice sheets are melting, contributing to sea level rise. The oceans have warmed and become more acidic as they have absorbed at least a quarter of the excess carbon dioxide being emitted by human activities.[14] Life in the world's oceans is threatened by these changes, compounding the threats caused by human overexploitation.[15]

Climate change is starting to threaten Earth's forests and destabilize the world's agricultural systems as well. Worsening droughts, heat waves, insect outbreaks, and wildfires have contributed to recent forest losses on every continent where trees grow.[16] A number of dramatic weather events—floods, droughts, and heat waves—have led to serious crop failures in recent years.[17] Recognition is growing among agricultural experts that climate change will soon threaten our ability to produce adequate food for the world's growing population.[18]

It will get worse before it gets better

Scientists have long known about lag time of several decades in the climate system's response to changes in greenhouse gas concentrations, but the implications of this are poorly appreciated. Greenhouse gases that we have already emitted lock us in to future warming even if we never burn a single additional ton of coal or cut a single additional acre of forest. The amount of warming in the climate pipeline—caused by the greenhouse gases we have already emitted—is estimated to be approximately double what we have experienced so far.[19] This additional warming will be accompanied by more frequent, destructive, and deadly weather disasters, particularly heat waves and floods,[20] as well as accelerating sea level rise,[21] ocean acidification, and all the other harms we're already feeling.

Much is uncertain

To this point, I've been emphasizing the strength of the scientific consensus about the causes, extent, and impacts

> The **Intergovernmental Panel on Climate Change** (IPCC) was established in 1988 by the World Meteorological Organization and the United Nations Environmental Programme to provide a scientific assessment of the state of knowledge about human-induced climate change. The IPCC issues a comprehensive report about every six years containing the latest information on the state of the climate (available on its website at www.ipcc.ch). Thousands of scientists from around the world contribute to the work of the IPCC, and all of its 195 member countries participate in reviewing its reports. As a result, these reports are consensus documents of the world's climate science community. The IPCC's Fifth Assessment Report, its most recent, was released in 2013 and 2014.

of climate change. Still, vital uncertainties persist. The biggest, most critical unknown concerns future warming: how much warming can we expect, and how fast will it happen? Scientists also have little confidence in predictions about components of the climate system that may cross thresholds and tip the climate into an abrupt or irreversible change.

Future warming is wildly uncertain. Figure 2, a graph from the IPCC's Fifth Assessment Report, shows global surface temperatures from 1850 to 2005 and projections for future warming under four different emissions scenarios up to 2100 and three of those scenarios from 2100 through 2300. The four greenhouse gas emission scenarios are a strong mitigation scenario (RCP2.6), two intermediate scenarios (RCP4.5 and RCP6.0), and a very high emissions scenario (RCP8.5). The lines after 2005 show the average estimates for each scenario, and the shaded areas

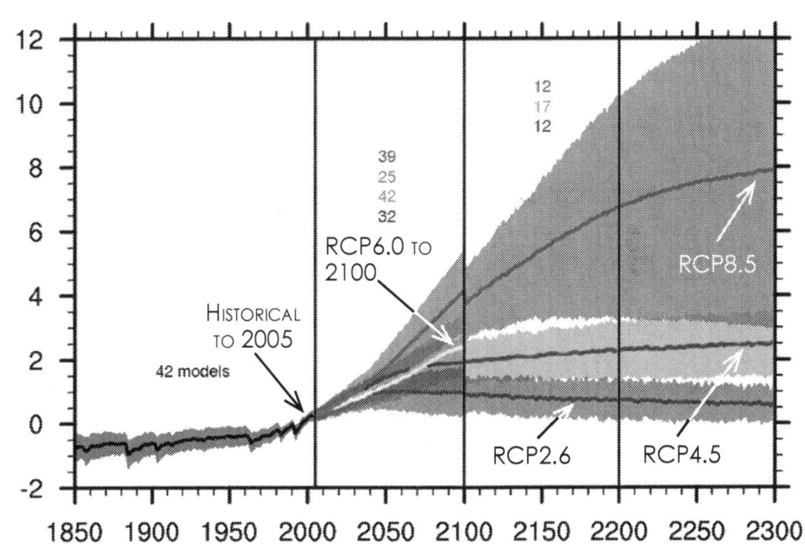

Figure 2. Global surface temperatures in degrees Celsius, actual from 1850 to 2005 and as projected under different scenarios from 2005 to 2300. Source: Figure 12.5 from Intergovernmental Panel on Climate Change, "Long-term Climate Change: Projections, Commitments and Irreversibility," Chapter 12 in *Climate Change 2013: The Physical Science Basis.*

represent the ranges of likely warming for each path. The numbers above the lines indicate the number of models used to calculate the mean predicted warming for each scenario during that time period.[22] The only things we can be pretty certain about are that warming will continue and that future greenhouse gas emissions will influence the amount of warming. Beyond that, we can say little with any certainty.

There are two major sources of uncertainty about future warming. The first is which emissions pathway humans will follow from here, and the second is exactly how much the climate will change in response to a specific change in greenhouse gas concentrations, something scientists call climate sensitivity. Climate sensitivity is measured in terms of how much a doubling of carbon dioxide concentrations will warm the world and is affected by feedback processes in the climate system that amplify or dampen the impact of greenhouse gases on the climate.[23] For example, when melting sea ice exposes dark ocean below, the ocean absorbs heat and amplifies warming; if cloud cover changes to reflect more energy back into space as temperatures rise, this slows warming.

Scientists still have only very rough estimates of climate sensitivity. The IPCC's Fifth Assessment Report places the climate's likely[24] response to a doubling of carbon dioxide concentrations in the range of a rise of 2.7 to 8.1 degrees F (1.5 to 4.5 degrees C).[25] A 2016 study re-examining the role of clouds in climate models raised the possibility that climate sensitivity may be up to 25 percent higher,[26] suggesting future warming could be even worse than projected in figure 2.

Taken together, the uncertainties about future human emissions and climate sensitivity result in an extremely wide range of potential future warming. According to the IPCC, temperatures may rise between 0.5 and 8.6 degrees F (0.3 and 4.8 degrees C) between now and 2100. It's difficult to communicate, based on these numbers, how critical these differences are. Today, all the effects we are experiencing are from the deceptively small-sounding 1.8 degrees F (1 degree C) rise in average temperatures over a century or so. An additional 1.8 degrees F (1 degree C) of warming in fewer than ninety years—the lower end of the projections, based on very rapid worldwide emissions reductions—will cause major disruptions. However, an additional 7 or 8 degrees F (4 degrees C) of warming in fewer than ninety years—the high end—would put our climate completely outside of human experience. This much warming would likely create an unrecognizably unstable climate, lead to massive sea level rise, vastly expand the extent of terrain hostile to modern civilization, cause frequent crop failures and consequent human starvation, and accelerate the extinction of many of the world's species.[27]

It looks bad

As the scientific consensus about climate change solidified, scientists and policy makers began to ask the question of how much additional warming human civilization could adapt to. In 2009, more than a hundred countries set the goal of limiting human-caused warming to 2 degrees C (3.6 degrees F) above the temperatures of preindustrial times,[28] approximately double the warming we've experienced since 1880. The IPCC's Fifth Assessment Report endorsed this 2-degrees-C limit in 2013.[29] In 2015, the Paris climate accord aspired to even less warming, with the goal of holding the increase to well below 2 degrees C and pursuing "efforts to limit temperature increase to

> There is nothing magical about a target of **2 degrees C**, and indeed, modern humans have never lived in a world that was 2 degrees C warmer than the recent pre-industrial climate. However, policy makers needed to set a goal, and this is where they set it based on available data. There is recent evidence that 2 degrees C of warming may be enough to trigger disastrous changes (such as the melting of the entire Greenland or West Antarctic ice sheet) or positive climate feedbacks (such as the release of huge stores of methane due to melting of large areas of permafrost) that could lead to runaway climate change.[31] The aspirational goal in Paris of limiting warming to 1.5 degrees C is a recognition of this sobering new evidence.

1.5 degrees C above pre-industrial levels."[30]

Even with all the scientific uncertainty, we know we are not on a path to limit emissions to contain warming to 1.5 or 2 degrees C and are unlikely to get there. The clearest summary I've found of how badly we are failing is in a 2014 paper in the scientific journal *Nature Geoscience* written by eleven of the world's top climate scientists.[32] Don't tune out here—this is important! According to this paper, limiting warming to 2 degrees C requires limiting total emissions (all human emissions over time) to a certain quota. *Two-thirds of the CO_2 emissions quota consistent with a 2-degrees-C limit had already been used up by 2014.* Given that global emissions of greenhouse gases were still rising when the paper was written (carbon dioxide emissions from fossil fuels and cement production grew 2.5 percent per year between 2004 and 2013), the authors concluded we were on the high end of IPCC emissions scenarios. Continuing on that path, we would use up the 2-degrees-C emissions quota in approximately thirty years.

Newer data suggests that global greenhouse gas

emissions may have leveled off in 2014 and early 2015,[33] opening the possibility that we may be veering off the high emissions pathway. However, on the high emissions path we were on in 2013, average temperatures would be likely to rise between 6.1 and 8.6 degrees F (3.4 and 4.8 degrees C) by 2100.[34] No one knows what a planet with this amount of warming would be like, but James Hansen has said it would be "a different planet." Immediate and sustained emissions reductions are the only known option to alter this scenario.

Given where we are today, and considering the time lag in the climate's response to emissions, the late and uncertain response of the world's political systems, and the inertia in the world's energy production and delivery systems, it's pretty hard to come up with any realistic scenario in which we do not overshoot the 2-degrees-C target and bring on a civilization-threatening climate crisis. In 2011, the International Energy Agency (IEA) examined the impact of continued construction of fossil-fuel power stations and infrastructure on future levels of warming. The IEA concluded that if the world didn't undertake a major shift away from this fossil fuel path immediately, making significant progress by 2017, we would be locked into "irreversible and potentially catastrophic climate change" of more than 3.5 degrees C (6.3 degrees F).[36] Keep in mind this statement came not from the Sierra Club but from the world's leading energy forecasting organization. Similar dire projections are contained in many other scientific analyses since 2010.[37]

This timeframe dovetails with what James Hansen began saying in 2006. At that time, Hansen estimated that we had about ten years to reverse worldwide greenhouse gas emission trends, and if we failed, disastrously large and rapid climate change would be essentially unstop-

James Hansen was an early pioneer in modeling Earth's climate. In 1988, as NASA's top climate scientist, Hansen gave the first scientific testimony to Congress on the evidence for and dangers of human-induced global warming. That testimony and his subsequent work established him as the leading scientific authority in the United States on climate change. He continued this work for twenty-five years as director of NASA's Goddard Institute for Space Studies. During George W. Bush's presidency, Hansen made headlines by speaking out about the administration's efforts to suppress the release of his research to the public.

As more years passed, Hansen became increasingly candid about the dangers of climate change and the urgent need to limit greenhouse gas emissions. He wrote a book about climate change for nonscientists and began giving talks and interviews. He was arrested in 2009 at a coal mine protest and in 2011 at a protest against the Keystone XL pipeline. In 2013, he retired from NASA to devote himself full-time to climate research, communication, and activism.[35]

pable.[38] Since then, Hansen has provided additional evidence of the need for immediate shifts off fossil fuels to avoid disastrous changes to Earth's climate.[39] Although reports from leading scientific institutions—the IPCC, the National Academy of Sciences, the American Academy for the Advancement of Science—have declined to use wording as strong as Hansen's, all have issued urgent warnings.[40] Yet globally, greenhouse gas emissions have not yet begun to fall.

What we do from here matters

The facts I've summarized here make it tempting to just give up and say it's too late. Because perhaps it is. I am not ignoring these facts when I say giving up is the wrong response and what we do from now on matters. I don't believe we can fix this or make it all go away, and I don't believe it's likely that in the next few years we humans will respond adequately to the challenge. Giving up is the wrong response because the future is unknown and the world we know is at stake, so if we can search for ways to make things better that is what we need to do.

I went to an event in 2011 called "The Eye of the Storm: Reimagining Ethics for a Changing Planet," where several hundred people gathered to talk about ways to respond to climate change. I found it by turns distressing, refreshing, heart wrenching, and inspiring, but what stayed with me most was something science fiction author Kim Stanley Robinson said: "Sometimes I get asked 'Is it too late?' *But this is the wrong question.* There are going to be losses. A better question is: 'Are we going to save less or are we going to save more?' I want to save more."

The basic scientific facts about climate change are settled, and the situation is dire—those are my starting premises. I'm also starting with the premise that we all want to save more.

Part 1 | The Signs of a New World

We may be, right now, at the pivotal moment when climate change begins to fundamentally disrupt human activities. Saying *now is the time when we come up against the planet's limits* certainly has a boy-who-cried-wolf feel about it. As I've just said, the future warming trajectory is highly uncertain. Yet even if we can't know the details, the fundamental picture is coming into focus. It now appears that climate change will be the last straw, rather than population growth, soil erosion, water pollution, toxic chemicals, or peak oil. But of course all these threats work together.

Until recently, I thought this reckoning might still be decades away, but now I think it will be sooner. If we are indeed at the start of a great climate disruption, we are living in a kind of suspended animation. In order to prepare, we need a better understanding of what we are bringing on. Yet most of us are still living our lives largely unconcerned about the urgent warning signs building around us. It's not always obvious what the relationship is between things like floods, wildfires, blizzards, or food prices and climate change. The next five chapters examine some of what is actually going on in our world today—the impacts of the planet's changing climate on weather extremes, other species, the oceans, Earth's forests, and agriculture. A sixth chapter addresses the puzzle of why so many of us are not noticing or caring as Earth's climate destabilizes all around us.

I suspect that whatever you know, the universal changes under way and their implications have not yet fully sunk in. Indeed, what I learned researching these chapters amazed me. The evidence suggests that in a few years we will look back and wonder how we could have missed all the signs that we now live in a new world.

1 Extreme Weather

I love big, dramatic weather: blizzards, thunderstorms, windstorms, even hailstorms and floods. I wasn't always like this, but I married a man who has always loved weather, and he definitely instilled this passion in our daughter. At some point, I realized I too had become involved. When a huge snowstorm followed by a deep freeze shut down Corvallis for more than a week one winter, when a thunderstorm boomed and flashed overhead for four solid hours one summer, when torrential rains brought flooding to our streets a few winters ago, we reveled. We don't want anyone to suffer, but we love seeing the power of nature show itself in the weather. Because of this passion, it's easy for me to remember that although 2011 was an unremarkable weather year in Oregon, it was a singular year for weather disasters in much of the rest of the country.

Weather disasters around the country—and the world

It used to be that in an average year, three or four major weather disasters—events that cause more than a billion dollars in damage—struck the United States. In 2011, according to the National Oceanic and Atmospheric Administration (NOAA), we experienced a record-breaking fourteen major weather disasters. These included a blizzard that dumped five inches to several feet of snow over twenty-two states; massive spring flooding on the Mississippi and Missouri Rivers; three extremely deadly spring tornado outbreaks; widespread drought and extreme heat across the Great Plains; a series of historic wildfires in Texas, New Mexico, and Arizona; and Hurricane Irene, which caused more than seven billion dollars in damage and at least forty-five deaths in the United States alone.[1]

I remember wondering if the year's weather might mark some kind of a turning point, when the connection between climate change and extreme weather would begin to crystallize in people's minds. I was reading about it everywhere; articles examining the relationship between extreme weather and climate change appeared in *Scientific American,* the *Guardian,* the *New York Times,* the *Los Angeles Times, USA Today,* the *Seattle Times,* and elsewhere. The IPCC issued a special report on managing the risks of extreme weather events with the changing climate.[2] As people reeled from each new weather calamity, the question most often asked was, are these deadly and destructive weather disasters *caused* by climate change?

The 2012 tally for major weather disasters was eleven, including the horrific Colorado firestorms and Hurricane Sandy.[3] In 2013, July brought the Arizona firestorm that claimed the lives of nineteen firefighters, and then Colorado's record-smashing floods in September covered an area the size of Connecticut; meteorologists struggling to depict the scale of the devastation described the flooding as biblical.[4] NOAA recorded eight separate billion-dollar weather disasters in the United States in 2014, including California's catastrophic drought and severe flooding in the Northeast.[5] In 2015, more records were broken.[6] South Carolina experienced catastrophic flooding when more than 20 inches of rain fell over three days, Boston received more than 7.5 feet of snow in twenty-three days, and snowpack in the Sierra Nevada hit an all-time low. We had the wettest month ever recorded across the contiguous United States in May, and in the summer, the most acres burned in a single wildfire season.[7] An analysis of federal data found that in the five years ending in September 2015, 96 percent of America's counties had been hit with a weather extreme.[8] The years of only three or four major weather disasters in the United States seem long gone.

At the same time, millions of people in other parts of the world have experienced record heat waves, floods, droughts, storms, wildfires, and crop failures in recent years, and the hottest year ever measured occurred in 2014, until 2015 broke the record again.[9] I'll refresh your memory with just a few examples. In 2010, a summer heat wave and wildfires in Russia killed more than ten thousand people and wiped out 40 percent of the country's grain crop.[10] Australia's weather has been like one long disaster movie. After a decade-long drought so persistent and severe people had begun referring to it simply as "the drying," a series of record floods struck beginning in December 2010. Horrific fires and the worst heat waves on record followed in 2012 and 2013.[11] When Typhoon Haiyan struck the Philippines on November 8, 2013, it was the most powerful tropical cyclone storm to make landfall in recorded history; some estimates placed the death toll as high as ten thousand people, with millions left homeless.[12] According to a recent United Nations report, the annual number of weather-related disasters around the world in the years 2005 through 2015 averaged 335, nearly double the number that occurred in the decade from 1985 through 1994.[13]

The science linking extreme weather to climate change

I'm still not sure if 2011 was a turning point where Earth's increasingly chaotic and extreme weather really got our attention. But something did change around that time; climate scientists began trying to explain the relationship between extreme weather and climate change in a more accurate and understandable way. While it's not correct to say climate change *caused* Hurricane Sandy or the Australian heat waves, scientists can say a lot about the relationship between extreme weather and climate change. Since 2011, they have become much better at communicating what they know.

The science linking certain extreme weather events—heat waves, floods, blizzards, and droughts—to climate change is quite strong and getting stronger. The most solid link of all is between climate change and heat waves. As Earth's *average* surface temperature has risen, this has shifted the odds toward record-breaking heat and away from record-breaking cold. Rare hot events are hotter and more common, and rare cold events less common.

Multiple studies have documented that heat waves are both more frequent and more severe than they were in the recent past.[14] In the United States, record-breaking cold was just as likely as record-breaking heat in the 1950s, but by the 2000s, record highs outnumbered record lows by a ratio of 2:1. Worldwide, the percentage of Earth's land surface experiencing extreme heat in summer skyrocketed from less than 1 percent before 1980 to 10 percent by 2011.[15] Although floods and storms make for dramatic TV images, it's heat that is the deadliest type of weather; extreme heat causes more deaths each year than hurricanes, tornadoes, floods, lightning, and earthquakes combined.[16] High-humidity heat waves are particularly deadly for the weak and elderly, while low-humidity heat waves contribute to the dry conditions that fuel wildfires. Extreme heat also kills or damages crops, contributing to crop failures.

The link between climate change and more severe rainstorms and blizzards is also well established. This link is not intuitive, but the fact is that warmer air holds more moisture than cooler air. The recent rise in average temperature has corresponded with an increase in moisture in the atmosphere, which is causing both more overall precipitation and a shift toward heavier, less frequent storms.[17] In the United States, total rain and snowfall has increased by about 7 percent in the last century, but the amount of rain falling in the heaviest 1 percent of rainstorms has increased 20 percent, leading to more frequent flooding.[18]

Across the planet, the frequency and severity of droughts has been increasing as well, and very dry areas around the world have doubled in size since the 1970s.[19] The factors that contribute to worsening droughts include shortfalls of rain and snow; earlier snowmelt; a shift away from light rains to rare, extremely heavy rains;

and increased evaporation due to higher temperatures. These factors have all been driven or worsened by climate change.[20]

Hurricanes, typhoons, and cyclones are different names for the same storm type—tropical cyclones—in different parts of the world. The relationship between these highly destructive tropical storms and climate change is not completely clear. In order to form, tropical cyclones require sea surface temperatures of at least 80 degrees F (27 degrees C), so the rise in sea surface temperatures associated with a warming climate should stack the deck in favor of more frequent storm formation. In addition, climate change is resulting in a warmer, stormier, moister environment, which generally favors higher-intensity tropical cyclones. However, these effects could be partly counteracted by increased shearing winds hindering the formation of storms.[21] The strongest tropical cyclones have increased in intensity in recent years, and evidence points to a combination of natural variability and climate change being responsible.[22] However, records are incomplete, making analysis difficult, and this has been an area of some scientific controversy.

For tornadoes, there is uncertainty as well. Climate reporter Justin Gillis of the *New York Times* explained the problem this way: "Weather statistics suggest that tornadoes are becoming more numerous as the climate warms. But tornadoes are small and hard to count, and scientists have little confidence in the accuracy of older data, which means they do not know whether to believe the apparent increase. Likewise, the computer programs they use to analyze and forecast the climate do not do a good job of representing events as small as tornadoes."[23] As a result, the science of how tornadoes are being affected by climate change is not yet settled.

Quantifying climate change's contribution

The new field of climate attribution science teases apart the relative contributions of natural variability and climate change to extreme weather events. Climate attribution employs a variety of methods, including examining climate models, running simulations, and analyzing weather data to calculate how much climate change has contributed to—or increased the probability of—specific extreme weather events. Climate attribution has critics who argue that this type of research oversimplifies the world and gives misleading results,[24] but I think it helps people conceptualize just how much climate change is contributing to extreme weather.

In a landmark paper published in *Nature* in 2004, scientists concluded that the chances of the record-breaking 2003 European heat wave, responsible for fourteen thousand deaths in France, were more than doubled by climate change.[25] In the years since that study was published, climate attribution research has been refined and, by most accounts, improved. Researchers have found that climate change played a role in some of the deadliest weather disasters of the last decade, including Hurricane Katrina,[26] the Pakistan floods of 2010,[27] and the heat waves in Russia (2010) and Texas and Oklahoma (2011).[28]

In a series of studies examining twelve major extreme weather events that occurred in 2012, scientists found evidence that climate change was a factor in half of them.[29] Similarly, for sixteen extreme weather events in 2013, climate change was found to have a significant role in eight of them.[30] A series of studies examining twenty-eight extreme weather events in 2014 found that, for a third year in a row, about half of the events had been made more likely or worsened by climate change.[31] Particularly dramatic is what researchers are finding about extreme heat: according to one recent highly regarded study, 75 percent of today's extremely hot days can be attributed to climate change.[32]

The bottom line: Loading the dice

I think the crazy weather of 2011 helped bring about a change—a good one—in language about extreme weather in the age of climate change. The question asked by journalists and weather reporters used to be "Was this weather disaster *caused* by climate change?" and the stock answer was "No single weather event can be attributed to climate change." But this was always the wrong question, way too simple for the complexity of Earth's climate system, so it often led to wrong or misleading answers.

These days, climate scientists often explain that adding greenhouse gases to the atmosphere is like loading the dice, increasing the chances of rare weather events. Thinking in terms of odds is important because in today's warming world, all weather, including extreme weather, is influenced by *the combination* of natural variability and anthropogenic climate change. It's deceptive to say "No single weather event can be attributed solely to global warming" unless you also say "No single weather event can be attributed solely to natural variability." As climate scientist Kenneth Trenberth of NCAR explained to journalist Alyson Kenward, "For extreme events, the question isn't, 'Is it global warming or natural variability?' It is always both. The question is just how much each is contributing."[33]

2 Life on Land

Growing up, I spent summers outside. My parents owned fifteen acres of golden grasslands and mixed-shrub-and-oak forest in the hills west of the Napa Valley wine country in California. The property had a modest rectangle of a house, a faded wood barn and chicken coop, some haphazardly fenced pasture and a seemingly random collection of ancient fruit trees. In my memory, it was a little enchanted; I'm pretty sure I was a little bit feral. We went barefoot and peed in the bushes like puppies. My favorite tree was a giant mission fig, standing squarely in the way between the gravel driveway and the house. My older sister and I would climb in there and stay, shaded and comfortable, dangling our legs from the low branches indefinitely. Wandering the property had its risks; the occasional warning rattle of a snake in the tall grass sent a tingle down my spine. For many years, my mom and we three girls spent summers there, my dad joining us on weekends.

These days I'm no longer impervious to the discomforts of the outdoors or the fear of deadly snakes. I'm not as quick to heal, and I wear shoes outside against the thorns. But something stays with me from those summers. I remain connected to the earth, the trees, bugs, grasses, and birds I spent so much time with as a child. When I feel bad, going outside always comforts me. And all the changes, the losses and the shifting of life as the climate warms feel real to me, like when you hold your hand to the ground and feel the vibration of a distant drum.

It may seem that the world is the same as it was then, but it isn't. Ask an ecologist. Our alteration of the atmosphere has set in motion changes to the ground rules for life. Across species and taxonomic groups, habitats and ecosystems, regions and continents, living things are experiencing swift changes, and living things are responding. Penguins, butterflies, polar bears, flying foxes, songbirds, pikas, frogs, insects, flowering plants, sea life: three and a half billion years of evolution are being tampered with, and it's miraculous and terrifying what's going on in response. The biologist in me is fascinated. The child in me is frightened.

Changes in distribution and seasonal activities

In 2002, I read an article about a quirky ornithologist named George Divoky who had been studying the nesting behavior of an obscure Arctic sea bird—the black guillemot—since the 1970s.[1] Black guillemots nest on Cooper Island, a tiny barren gravel and rock island in northern Alaska; in his decades of studying them there, Divoky had accidentally stumbled upon and documented one of the first carefully observed examples of an organism responding to climate change. Year by year, Divoky watched as the birds' nesting date advanced earlier each spring, responding to earlier snowmelt and earlier arrival of the Arctic spring. The guillemots' breeding season grew longer each year, giving them more time to raise their chicks, and Divorky documented the rapid growth of the Cooper Island guillemot population as the Arctic warmed.

In the years since I read about Divoky's guillemots, biologists have documented changes in the distributions of plants and animals and in the timing of seasonal events in all well-studied terrestrial, marine, and aquatic systems, across taxonomic groups, and around the world. Many species' ranges are moving poleward and up in elevation, and seasonal events such as bud burst and nest building are advancing earlier. Many insect pests and fungal pathogens are spreading into new regions as climates shift. Some organisms are in severe decline from direct or indirect effects related to warming: in a few cases, extinctions of populations and even species have been attributed to climate change.[2]

A 2003 examination of worldwide studies of 1,598 species found that 59 percent of them showed measurable changes in seasonal activities or distributions. Ranges had shifted toward higher latitudes, and for species along elevational gradients, to higher elevations.[3] A 2011 meta-analysis—a study that analyzes data from multiple previous studies—found that organisms had moved an average of 16.9 km per decade poleward, and 11.0 meters per decade to higher elevations, two to three times faster than reported in earlier studies.[4]

Butterflies, whose first emergence in spring is often correlated with temperature, are emerging earlier in the United Kingdom, Spain, central California, and elsewhere, and have expanded their ranges northward across Europe. Lowland birds began breeding up in mountain cloud forests in Costa Rica in 1985. Swiss alpine flora has been expanding toward the peaks since the 1940s. Mangrove forests are spreading north along the Florida shoreline.[5]

Studies have found that many birds are laying their first eggs much earlier in the spring as warming progresses.[6]

Sometimes, these shifts throw an organism out of synchrony with its host, prey, pollinator, or critical food source. Yellow-bellied marmots, high-elevation herbivores that hibernate for about eight months each year, provide just one example. Researchers in Colorado found that over a twenty-four-year period, marmots were emerging twenty-three days earlier in spring, which correlated with a rise in local temperature of 1.4 degrees C (2.5 degrees F). However, over the same period, no change was seen in the date of spring snowmelt or plant flowering.[7] Butterfly-host asynchrony, where insect emergence and blooming of nectar sources have responded differently to local climate change, is threatening the viability of multiple butterfly populations.[8]

Winners and losers

Some impacts of climate change on living things are dramatic, and even gruesome. Pictures of drowned polar bears have become powerful symbols of climate change's reach. Along the bears' southern range boundary, they are declining in numbers and losing weight as the longer ice-free periods on Hudson Bay cause them to starve. Warming has also been linked to declines of the bears' main food, the ringed seal.[9] Effects of climate change—increases in heavy rains, stronger storms, and heat waves—are killing baby penguins in Argentina and peregrine falcon chicks in the Canadian Arctic.[10]

More than half a world away in the subtropical forests of New South Wales, Australia, flying foxes have been dying from extreme heat. During a record heat wave in January 2002 (when the temperature reached 109 degrees F at the nearest weather station), biologists Justin Welbergen and Stefan Klose documented thousands of flying foxes falling from the trees where they doze during the day, most of them dead from the heat.[11] Since that initial report, Welbergen and colleagues have documented multiple extreme heat waves killing tens of thousands of flying foxes in Australia, including a single day in January 2014 when more than 45,500 animals died from extreme heat.[12] Welbergen's analysis of temperatures near flying-fox colonies found that temperature extremes lethal to flying foxes have been increasing since 1950.[13]

The changing climate is rippling through living systems in other ways as well. Historically, whitebark pines have grown in climates where it's too cold for mountain pine beetles to overwinter, so this species has not been subject to serious beetle outbreaks. With climate change, the bitter cold snaps that kill overwintering beetles are disappearing, so beetle populations are killing previously protected whitebark pine forests.[14] One recent study found that warming has resulted in a doubling of the mountain pine beetle's season, with exponential increase in beetle populations of up to sixty times as many beetles in any given year.[15]

Occasionally, a species benefits from the changing climate. Trumpeter swans, North America's largest species of waterfowl, were nearly hunted to extinction more than a hundred years ago. These birds breed in forested ponds and lakes in interior Alaska and then migrate more than a thousand miles south to winter in British Columbia, Washington, and Oregon. Since the 1960s, rapid warming in Alaska has lengthened their breeding season, expanded their summer range to the north, and increased the abundance of their food: as a result, swan populations have been rising steadily.[16] Chinstrap and gentoo penguins, which feed in the open ocean, have benefited from the retreat of Arctic sea ice and expanded southward along the Antarctic Peninsula over the past twenty to fifty years.[17] The gray whale also appears to be benefiting from the rapid melting of Arctic sea ice and has expanded its range north into the Chukchi Sea as more and more open ocean is available during summer.[18]

While climate change is creating a few winners, it's creating far more losers. Even as open-ocean-loving species have thrived, sea-ice dependent penguins have suffered. On the Antarctic Peninsula, northern Adélie penguin populations have fallen by 90 percent, and the ice-dependent emperor penguin colony is now extinct.[19] In the Central and South American tropics, amphibians have been disappearing precipitously. One study found that 67 percent of harlequin frogs in this region had disappeared over only two to three decades, and fungal pathogen outbreaks associated with warming are believed to be largely responsible.[20] In Spain, butterfly species lost on average a third of their distribution area over a thirty-five-year period. In North America, entire populations of Edith's checkerspot butterfly have gone extinct while the remaining populations have shifted northward.[21] Many North American and European bumblebees are disappearing from the southern portions of their ranges while failing to move north,[22] a disturbing trend that could harm crops, ecosystems, and people as bumblebees pollinate many farmed and wild plants throughout their ranges.

Mountaintop and polar species are particularly at risk, because the climate is warming faster at the poles and at high elevations than elsewhere, and these species have nowhere to migrate to as their habitats warm. Extinctions of pika populations along the lower elevational boundaries of their range in the western United States have been linked to warmer summer temperatures, less insulating snow in winter, and less precipitation, suggesting a highly uncertain future for this mountain-dwelling mammal.[23] The rapidly warming Arctic is seeing declines in many of its bird species, including murres, peregrine falcons, ivory gulls, and long-tailed jaegers.[24]

A worldwide, multifaceted selection event

Our alteration of the atmosphere has set in motion a worldwide, multifaceted selection event. Some species will move, adapt, and survive, but many won't. Habitat loss, pollution, and other human stressors are already contributing to a sharp rise in the rate of extinctions, which is currently estimated to be a hundred to a thousand times higher than the background rate.[25] Climate change, when added on top of other human threats, is projected to dramatically increase species extinctions. One study warns that an additional tenfold rise in the global extinction rate is possible by the end of this century.[26] Another analysis predicts that if current warming trends continue, one in six species may be driven to extinction.[27]

Short-lived widespread species (for example, insects and mice) will generally have better chances than long-lived specialists with small ranges (such as grizzly bears and redwood trees). In addition, indirect or second-order impacts of the warming climate will devastate some previously healthy species (as illustrated by the rapid crashes of harlequin frogs and whitebark pines). No one knows how far the changes will go and what kind of living communities will be left if rapid warming proceeds on its current trajectory. But what has happened to the black guillemots in George Divoky's study shows just how unpredictable the impacts of a changing climate on living communities can be.

In the four decades that Divoky has been studying black guillemots on Cooper Island, their fortunes have changed dramatically, twice. These birds need about eighty snow-free days to breed successfully. From Divoky's arrival on the island in 1975 to 1995, he documented that the snow melted five days earlier per decade, allowing the birds to nest earlier and raise more chicks. During those years, Divoky watched as the guillemot population expanded and thrived. This was the change in their fortunes I read about in the 2002 article.

But black guillemots spend their entire lives in the Arctic, feeding at the edge of the pack ice. As the Arctic continued to warm, the pack ice retreated farther offshore each summer; at a certain point the guillemots began to have difficulty reaching the ice from their nests to feed. In the mid-1990s, some chicks began to starve, and birds began to die off. Then, a second threat emerged: as polar bear prey began declining with warming, polar bears started raiding the guillemot nests and eating chicks, something they had never done before. Although Divoky responded by installing bearproof nest boxes, he believes that ultimately climate change—which initially helped the Cooper Island guillemots to thrive—may lead to the whole colony shutting down.[28] Assuming he is able to continue his annual research for long enough, Divoky will have been there to witness the whole thing.

3 Oceans

One November weekday when the weather forecast looked hopeful, I took off for a day at the coast. I headed to Seal Rock, a protected cove about an hour's drive from home. Before Lia was old enough for school I took her to the Oregon coast often; it was a reliable escape from the sometimes-desperate tedium of keeping a toddler occupied, and we both loved the ocean. We would spend the entire day digging in the sand and visiting the aquarium, where she'd run from one exhibit to another, barely slowing to take in the unusual colors and shapes of the amazing marine life. I remembered from those years that Seal Rock's beach was protected from the frequently strong winds that can make the Oregon coast such a challenge. The tide was high when I arrived, reducing the sandy crescent at the end of the steep trail to a narrow strip that was barely wide enough to walk. But the sun was out, I was blissfully alone, and as I had remembered, the winds were cut down to gentle breezes. I set up my chair on a slope overlooking the beach. From my perch, I waited to see if the tide was turning and breathed in the ocean air.

Looking out at the vast Pacific, I thought of Julia Whitty, an environmental journalist and author I once met who wrote a book about the ecology of the world's oceans. In an article she wrote about the threats these great bodies of saltwater face, she says, "The ocean is our blind spot: a deep, dark, distant and complex realm covering 70.8 percent of Earth's surface. We have better maps of the surface of Mars than of our own sea floor."[1] Sitting there, on the edge of the largest of Earth's connected oceans, I wondered how it's possible to have a blind spot for something so huge. Yet I think Whitty is right that we don't really see the ocean's dominating presence on the planet or understand much about it; we don't realize its central importance to our atmosphere, our food systems, our weather patterns, water cycles, our very existence. Even for me, a self-professed nature lover, so much about the ocean is unknown, mysterious, or forgotten.

Here are a few amazing facts about Earth's oceans. They contain 97 percent of the water on the planet; of the rest, 1 percent is freshwater and 2 percent is ice.[2] The earliest life evolved in the oceans and predates life on land by *three billion years*. One in every seven people depends on food from the oceans as their primary source of protein. Humans also depend heavily on phytoplankton, the tiny photosynthesizers of the oceans that make up half the plant matter on Earth and produce half the oxygen in our atmosphere.[3] And then there is the global system of ocean currents—sometimes called the great ocean conveyer belt—that circulates seawater around the world. This conveyer belt operates because of density differences in water at different temperatures and salinities. Cold salty water is dense and sinks, while warm freshwater rises. It takes about a thousand years for this moving water to complete one full circuit around the globe.[4] This planetwide ocean current system stores and transfers heat, playing a critical role in shaping Earth's climate.[5]

As vast as the oceans are, we are currently changing them and the life they harbor in dramatic ways. Overfishing, invasive species, sewage and fertilizer runoff, plastic pollution, and soil erosion have been altering the oceans for many decades. Now climate change is creating widespread upheaval by altering ocean temperatures, changing seawater's pH and salinity, melting ice, and accelerating sea level rise; these changes are precipitating a host of other shifts to living systems and to Earth's climate.

Direct effects of climate change on temperature and pH

Climate change has two direct effects on the oceans: temperatures go up, and pH goes down. Since the 1950s, upper ocean heat content has risen as the oceans have absorbed an estimated 90 percent of the added heat trapped in Earth's atmosphere.[6] The mean sea surface temperature (SST) has increased at least 0.4 degree C (0.7 degree F) since the 1950s, with this increase unevenly distributed across the world's oceans.[7] A 2014 study suggests that warming of the worlds' oceans may actually be significantly greater than this estimate due to spotty sampling of oceans in the Southern Hemisphere.[8]

Meanwhile, the oceans have also been absorbing an estimated one fourth of the CO_2 that human activities are adding to the atmosphere. Climate scientists once considered the ocean's absorption of CO_2 a great service, but it has serious downsides.[9] Rising ocean uptake of CO_2 produces carbonic acid, which increases the ocean's acidity. Over the past 150 years, sea surface pH has dropped by an average of 0.1 pH units, an increase in acidity of

26 percent. Most of this acidification has occurred in the last few decades.[10]

Rising SSTs and ocean acidification have been associated with a cascade of other dramatic shifts. They include increased stratification (or vertical layering) in the oceans, altered patterns of ocean circulation, changes in precipitation patterns, and altered freshwater input. Other changes include reduced subsurface oxygen concentrations and changes in salinity. In the upper ocean, low salinity regions are getting fresher, while high salinity regions are getting saltier.[11] Finally, as water warms, it expands, and this thermal expansion contributes to sea level rise.[12]

Upheaval for life in the seas

Marine scientists have documented that changes in temperature, circulation patterns, stratification, oxygen content, and ocean pH caused by global warming are throwing sea life into upheaval—from plankton to whales and from the equator to the poles.

As ocean conditions change, organisms' tolerances for temperature, oxygen levels, and pH are being tested and sometimes exceeded, not unlike what has been happening on land. A comprehensive survey of all available studies on changes in marine organisms found species are migrating toward the poles; altering their breeding, feeding, and migration patterns; and responding rapidly to the altered conditions in the oceans brought about by climate change.[13] Marine organisms are changing their dispersal patterns and the timing of reproduction or migrations. Many marine fish and invertebrates are moving to higher latitudes and greater ocean depths with warming. Sometimes species are living together that have never before interacted with one another. The most dramatic changes are occurring in the tropics and at the poles, but life in all marine regions is being affected by climate change.[14]

Food webs are fraying in many ocean ecosystems, and climate change is a large contributor. Ocean food webs depend on zooplankton and phytoplankton—tiny animals and plants—whose abundances fluctuate naturally. However, as ocean surface waters have warmed, oxygen levels have fallen, causing declines in plankton and other prey. Food web destabilization is believed to be contributing to shrinking salmon runs in the Pacific Northwest,[15] as well as recent crashes of North Atlantic right whales dependent on the food web that includes zooplankton, cod, mackerel, herring, and haddock.[16]

Tropical coral reefs and the species that depend on them are threatened by both rising temperatures and ocean acidification associated with climate change.[17] Corals are made up of two organisms—coral and algae—living symbiotically as one. They are sensitive to small increases in water temperature, which can cause something called coral bleaching—when corals expel their algae, losing their color and their ability to photosynthesize, which provides the vast majority of their nutrition. Although coral can sometimes recover from bleaching events, recovery takes decades.[18] As ocean temperatures rise, bleaching events are becoming more frequent and severe, weakening and killing reefs.[19] In 2015, the third global coral bleaching event on record began (the first was in 1998, and the second in 2010).[20]

Ocean acidification poses a second serious threat to coral, decreasing the saturation levels in seawater of the calcium carbonate ions corals use to make shells. This raises the energetic costs of building reefs, slows coral growth rates, and reduces coral's ability to recover from stress of all kinds.[21]

The implications of coral's susceptibility to global warming are not pretty. Coral reef systems are among the most vulnerable and the most valuable for other species of any ecosystems on the planet. They are especially vulnerable because coral reefs can live only in specific places, and are already threatened by overfishing, pollution, sedimentation, nutrient enrichment (from sewage and fertilizer runoff), and invasive species. Coral cover has already been reduced by an estimated 80 percent in the Caribbean and 50 percent in the Pacific.[22] Their value is not easily measured, but consider this: coral reefs are among the most biodiverse ecosystems on Earth, taking up only about 1 percent of the space in the marine environment but associated with an estimated 25 percent of all marine species.[23] It is estimated that 30 million local fishermen and women around the world rely on coral reefs for their livelihoods.[24]

Many other marine organisms—sponges, marine worms, clams and oysters, starfish and sea urchins, crustaceans such as crabs and lobsters, and some types of plankton—also build shells or skeletons from calcium carbonate dissolved in seawater. All of these organisms are endangered by ocean acidification as well. Ken Caldeira, a leading expert on ocean acidification, modeled future ocean pH levels under different future emissions paths and found that under all of them, ocean acidity would at least double from preindustrial levels by the end of this century; his conclusion was that any future in which we do not quickly bring our carbon dioxide emissions down to levels approaching zero threatens to cause the extinction of the entire category of organisms that build calcium carbonate shells.[25]

The effects of melting ice

Polar regions are warming quickly, and as they warm, ice melts. All this melting ice is affecting ecosystems, species, sea level, and the climate system itself. It wasn't until I began writing this chapter that I fully grasped that the ice melting at the poles is actually two very separate phenomena, each with profound, but distinct, implications. Some of the ice is in ice sheets, and some in sea ice.

Greenland in the north and Antarctica in the south are landmasses that are largely covered by ice sheets. When ice sheets melt, the melted water is added to the oceans, directly causing sea level rise.[26] According to NASA, ice sheets at both poles are melting rapidly, the rate of loss has increased dramatically, and ice sheets are now losing more than three times as much ice each year as they were in the 1990s. So far, about two thirds of this ice loss comes from the Greenland ice sheet, and overall, scientists estimate that ice sheet melting is responsible for only about a fifth of the global sea level rise observed since the 1990s.[27] However, several 2015 studies have raised alarms about the potential for runaway ice sheet melting that would lead to catastrophic levels (ten feet or more) of sea level rise.[28]

Sea ice is composed of frozen seawater floating in the ocean. Arctic sea ice cover expands during winter months and partially melts during summer, usually reaching its annual minimum in September or October. Therefore, sea ice needs to be measured at the same time each year for annual comparisons to be valid. Unlike the melting of ice sheets, melting sea ice has little direct effect on sea level; think of melting ice cubes in a glass of water. However, sea ice is of vital importance to humans and animals in polar regions. Sea ice also plays a key role in regulating Earth's climate.[29]

Year to year, sea ice cover fluctuates due to weather variability and ocean conditions, but the overall trends for Arctic sea ice are clear. Figure 3 charts the extent of September Arctic sea ice from 1979 to 2015. Since 1979, Arctic September sea ice extent has declined approximately 13.4 percent per decade.[30] Arctic sea ice is also getting thinner, making it more vulnerable to melting.[31] Extrapolating from current melt rates suggests that late summer sea ice may disappear from the Arctic by the 2030s.[32]

Climate change deniers often bring up the fact that in the Southern Hemisphere sea ice has been increasing, and technically this is true. However, Antarctic sea ice expansion has been small, increasing by 1.1 percent per decade compared to the 1981 to 2010 average. Although the reasons for sea ice expansion in the Southern Hemisphere are uncertain, many experts believe that this small expansion is due to changes in wind and ocean circulation patterns.[33]

I've never seen sea ice, which makes it harder to fully appreciate its importance. But for the seabirds, mammals, and people who live and hunt on the ice, its loss takes away their habitat and threatens their survival. One of the most vulnerable species is the polar bear, because it hunts primarily on sea ice. Seals, penguins, and many Arctic seabirds have fates tied closely to sea ice as well.[34] The long-term study of black guillemots discussed in Chapter 2 provides a detailed illustration of how the fabric of entire Arctic communities is beginning to unravel as the sea ice recedes.

Sea ice also influences the global climate, having an overall cooling effect. The bright white surface of sea ice reflects much of the sunlight that strikes it back into space,

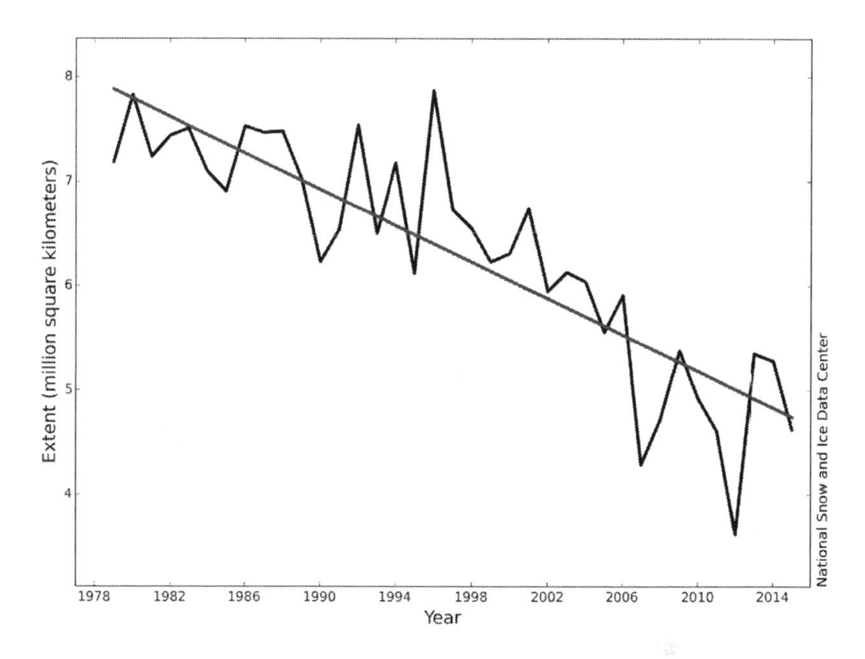

Figure 3. Average monthly Arctic sea ice extent, September 1979–2015. Source: Figure 3, "2015 Melt Season in Review," Arctic Sea Ice News and Analysis, National Snow and Ice Data Center, October 6, 2015.

the same way white paint on a building reflects sunlight and keeps it cool. When sea ice melts, the dark ocean surface that replaces the ice absorbs sunlight, causing further warming, leading to more melting. Therefore, the loss of Arctic sea ice creates a positive feedback system that increases global warming. Sea ice also affects ocean currents, contributing to the movement of the great ocean conveyer belt.[35]

Rising seas: How much, how fast?

The melting of ice—from both glaciers and ice sheets—and thermal expansion of water as it warms are the two primary contributors to rising sea levels, and both these processes are accelerating due to climate change.[36] Since the late 1800s, global sea level has risen, on average, about 8 inches.[37] Satellite measurements suggest that since 1992, the rate of sea level rise has roughly doubled from the rate observed over the last century.[38]

Projecting future sea level rise is one of the most uncertain of all climate predictions. The latest IPCC report predicts that sea level is likely to rise between 10 and 39 inches (or about 1 to 3.2 feet) by 2100.[39] However, current climate models cannot take rapid ice sheet changes into account,[40] so ice sheet contributions to future sea level rise are largely unclear. The Third National Climate Assessment estimates that average sea level is likely to rise between 1 and 4 feet by 2100, but also suggests that given the uncertainty, decision makers may want to use a wider range of scenarios for planning purposes, from 8 inches to 6.6 feet.[41] Other analyses suggest the lower end estimates are unrealistic; several studies have posited sea level could rise between 1.6 and 6.6 feet by the end of this century.[42]

Even with this uncertainty, we can expect that the level of damage and disruption caused by sea level rise this century will be stunning. Most of the world's major cities are located along coastlines. Sea level rise will cause direct coastal flooding, but it will also dramatically increase the odds of highly destructive flooding from storm surges,[43] such as occurred with Hurricane Sandy. In the United States, nearly 5 million people live less than 4 feet above high tide line, half of them in Florida. Locally, the amount of sea level rise varies, based on such things as ocean circulation patterns and tectonic history. The land along the East Coast is still subsiding after the last ice age, resulting in faster sea level rise.[44] Another global trouble spot will be Bangladesh, most of which lies close to sea level and is already prone to flooding from cyclones.[45] Low-lying populations everywhere face similar threats; it is estimated that 150 million people around the world live 3 feet or fewer above sea level.[46]

Messing with the great ocean conveyer belt

The great ocean conveyer belt redistributes vast quantities of heat around the planet, and it has a profound influence on Earth's climate, as well as ocean nutrient and carbon dioxide cycles.[47] Europe's climate is strongly buffered by the great ocean conveyer belt; Norway has an average temperature almost 20 degrees F (11 degrees C) warmer than that of Manitoba, although they are located at similar latitudes.[48] And the conveyor belt helps drive the vast warm Gulf Stream current,[49] with multiple impacts on the climate of the eastern United States.

Increased precipitation and widespread melting of ice—both processes that are expected to continue as the climate warms—expand the upper layer of freshwater in the oceans, which slows and could eventually shut down the great ocean conveyer belt. Researchers have predicted that significant additional warming (3 to 5 degrees C or 5 to 9 degrees F) would be required to shut down the great ocean conveyer belt,[50] but evidence exists that warming may already be weakening the massive ocean currents by 15 to 20 percent.[51] A shutdown would cause rapid and highly dramatic regional climate shifts. For instance, Europe would likely become much colder and experience continent-wide crop failures.

The day I spent at Seal Rock, I didn't actually worry about all this. I had been learning that all our ocean measuring and modeling has left us with huge worries and many unknowns. I understood the bleak evidence that climate change is adding powerful threats to Earth's already over-exploited and stressed oceans. I believed Julia Whitty, who warns, "In our modern ocean, beset by dwindling phytoplankton, warming waters, melting ice, rising acidity, corroding reefs, dying shellfish beds, and collapsing food webs, life itself threatens to sputter out."[52] I knew of no solutions on the scale of what is needed. Yet I came home tired, content, and refreshed, the way I always feel after spending a day at the ocean's edge, having breathed in the winds from the seemingly infinite Pacific. Feeling this way defies logic, knowing what I know about the state of the oceans. I'm starting to see that coming to terms with climate change requires developing an understanding that is completely distinct from logic, or at least has a type of logic all its own.

4 Forests and Trees

One day, I came across a passionate essay about trees by Montana journalist Jim Robbins in the *New York Times*. He said, "We have underestimated the importance of trees. They are not merely pleasant sources of shade but a potentially major answer to some of our most pressing environmental problems. We take them for granted, but they are a near miracle."[1]

Like many Westerners, I crave wide-open views and vast skies. When I moved from California to the East Coast at eighteen, I remember going hiking in the tight-growing forests of New England and feeling disappointed, and slightly claustrophobic. With time I grew to appreciate the famous fall colors of Vermont and to see that the mixed hardwood forests possessed their own kind of beauty. Yet moving home again, I adored getting up on the unobstructed ridgelines. Yes, I do love Western conifer forests. Still, even after many years in Oregon, where the symbol on our license plates is a stately evergreen tree, I always seek clearings and places above the trees where I can see across the land.

Perhaps I do not appreciate trees enough. Because trees *are* a near miracle. They convert sunlight into food, habitat, and wood. They filter and clean up water, soil, and air; they make shade and moisture that cools the climate and holds soil in place. And they absorb and store vast amounts of carbon dioxide. However, perhaps because we can't bring them into the lab to study, trees are also a near mystery. Researchers are still fuzzy about much of how trees do what they do and how forests function.

The role of forests in carbon absorption

It was in the 1950s that a scientist named Charles David Keeling figured out how to accurately measure carbon dioxide concentrations in the atmosphere. This discovery led to Keeling's continuous measurements of atmospheric CO_2 that today are referred to as the Keeling Curve. Keeling's method became a critical piece in understanding climate change,[2] yet his discovery soon uncovered a mystery. Carbon dioxide levels had been increasing, but only by about half as much as scientists calculated they should have been given the amount of carbon dioxide human activities were releasing. For many years, the question of where the missing carbon dioxide was going remained unanswered.

According to scientists' best estimates, the world's oceans are absorbing about half of the missing carbon, and the world's forests are absorbing the other half.[3] Several things are especially impressive about the big role of forests in carbon absorption. One is that while the oceans cover approximately 70 percent of Earth's surface, forests cover only about 9 percent.[4] The other thing that makes trees a near miracle is that absorbing carbon dioxide doesn't harm trees. In fact, higher carbon dioxide levels can sometimes enhance tree growth, although this effect has often been overestimated.[5]

In every patch of forest, all over the world, trees absorb carbon dioxide as they grow and release carbon dioxide as they go dormant or die. On balance, living forests typically absorb more carbon than they release, sucking up carbon and acting as carbon sinks. However, under conditions of widespread tree death, fire, or deforestation, carbon stored in tree tissue is emitted into the atmosphere as they decay or burn, so dying forests become carbon sources. Worldwide, the balance of forest growth and forest death is what largely determines forests' overall impact on carbon dioxide levels in the atmosphere.

Estimating the amount of carbon dioxide absorbed, stored, and released by the world's forests is a complex challenge. It requires accurately measuring the absorption and release of carbon in enough forest types and subtypes to get a global picture. Forests are classified into three major types, according to latitude: tropical, temperate, and boreal. Tropical and temperate forests are then subdivided into groups based on seasonal distribution of rainfall.[6] According to a 2011 analysis, it's the world's temperate and boreal forests that are absorbing roughly a quarter of the carbon dioxide released by humans each year. Tropical forests have little overall impact, because according to most estimates, tropical deforestation has largely canceled out the carbon-absorption capacity of tropical forests in recent decades.[7] Arresting tropical deforestation could allow tropical forests to absorb carbon and lessen climate change—more on this in Chapter 12.

How climate stress affects forests

Unfortunately, evidence suggests that the warming climate increasingly threatens trees in all regions of the world. In 2013, several prominent forest researchers published a review of the existing literature on how climate stress is affecting forests. They found that forest die-offs from drought and heat stress have occurred on every continent except Antarctica, which has no trees. They predicted that forest die-offs will increase with climate change, threatening the ability of Earth's forests to absorb and store carbon.[8] In 2015, a series of papers in the journal *Science* assessed the condition of forests around the globe. These papers documented the decline of forest health and predicted that the combination of climate change and human consumption of land and trees will lead to the continued deterioration of Earth's forests.[9]

Here are a few examples of the types of forest changes documented in recent years. Two mega droughts in the Amazon basin, in 2005 and again in 2010, killed billions of trees and led to large-scale releases of carbon.[10] Another study found that the Amazon rainforest is absorbing carbon at a slower rate today than in the 1990s, reducing its capacity to act as a carbon sink.[11] In Australia, eucalyptus forests are dying from heat, and in South Africa, euphorbia trees are succumbing to heat and water stress.[12] Boreal forests form a vast ring around the upper Northern Hemisphere and make up 30 percent of Earth's forest cover. But this region is warming twice as fast as the rest of the world, and huge tracts of boreal forests are dying from fires, insect outbreaks, and other stresses.[13]

In the western United States, views of dead and dying forests are widespread and dramatic. I've seen some of it, in Oregon and Colorado, and I've seen pictures of other places. But the scale of forest death, the variety of forest types being affected, is something few of us grasp. Scientists surveying western forests in the early twenty-first century across elevations, regions, and species found that the death rates of trees had doubled over the past two to three decades across the western United States. Tree mortality was up in every tree type, at all elevations, for all tree sizes, and in every region. These scientists concluded that regional warming and drought stress were the most likely causes of surging tree mortality rates.[14]

Colorado is losing its iconic quaking aspen forests from drought stress.[15] Ancient bristlecone pine forests are dying too, and the 2011 drought in Texas killed more than half a billion trees in cities, parks, and forests.[16] The whitebark pine has dominated high-elevation and high-latitude forests over much of western North America for centuries; today it is estimated that at least half of all whitebark pine trees are dead or dying.[17] With the deep freezes that destroy overwintering pine beetles disappearing, beetle populations are expected to keep growing and killing whitebark pine forests.[18]

Describing the scale of current forest losses in North America, Jim Robbins writes: "Across western North America, from Mexico to Alaska, forest die-off is occurring on an extraordinary scale ... All told, the Rocky Mountains in Canada and the United States have seen nearly 70,000 square miles of forest—an area the size of Washington State—die since 2000."[19]

Fire season: Longer and hotter

And then, of course, there are forest fires. In 2006, I read an article in *Science* about wildfires in the western United States that stunned me. It examined the frequency and intensity of wildfires since 1970. The analysis found that wildfire activity had increased suddenly and dramatically in the mid-1980s, with more frequent, larger, longer-lasting wildfires and longer fire seasons. Overall, the authors found a fourfold increase in major wildfires in the West, a sixfold increase in forestland burned, and a seventy-eight-day lengthening of the fire season associated with the warming conditions.[20]

These same findings were confirmed and updated in a 2012 analysis of forty-two years of Forest Service data. This second analysis found that western U.S. forests in the 2000s had seven times more large fires than during the 1970s, fire season had lengthened by seventy-five days, and wildfires burned twice as much land each summer as they did in the 1970s.[21] Here in the central Oregon Cascades, 43 percent of the Sisters Ranger District burned between 2002 and 2012.[22]

Across the American West, people have fought bitter fights over wildfires and forest management for decades. Forest ecologists teach that most western forests evolved with wildfire and that fire suppression itself is an important contributor to the catastrophic wildfires we've seen in recent years. In these systems, a hundred years of fire suppression has allowed dangerous fuel loads to build up, setting the stage for uncontrollable and highly destructive fires. As a remedy, forest ecologists have proposed controlled burns and, in some areas, "let it burn" policies to clear away dangerous fuel loads and restore forests to a more natural state. This was the paradigm I was taught in graduate school. On the other side, many trained foresters and those in the forest industry believe that restrictions on logging, firefighting, thinning, and replanting have left forests unhealthy and vulnerable to fire. They propose logging and thinning forests wherever possible, and fighting all wildfires aggressively from the start. This debate still takes place in communities across the West, although I suspect it is increasingly irrelevant.

It's not that the things we argue about—firefighting, logging, thinning, forest management, and building houses in forests—have no effect on wildfires. But the evidence suggests that the impacts of climate change are rapidly overwhelming all these factors. The truth on the ground is that the rules of forest fires in the West have changed. You can see it in central Oregon, in the Colorado Rockies, in Yosemite, in Arizona, and even in Alaska. In California, the fire season has become essentially year-round. And this is only the beginning. A National Research Council report projects that for each 1 degree C (1.8 degrees F) increase in average temperature, the area burned in the western United States could quadruple.[23]

The silence about these unbelievably rapid changes in fire behavior has been a loud one, until very recently. Today it is finally being recognized that western forests are burning as the climate warms.[24] According to a 2015 report, firefighting now takes up more than 50 percent of the U.S. Forest Service's annual budget, up from 16 percent a decade ago. In ten years, it could consume three quarters of its budget.[25] For me, the evidence of this rapid change brings up the question, what if neither controlled burns, nor logging and thinning, nor firefighting will bring these forests back? What if it really doesn't matter that much what we do?

Although wildfires in the West are the most vivid to me, the rest of the world's forests are susceptible, too. According to a 2015 report, the number of large wildfires in the Arctic has increased nearly tenfold in the last fifty years.[26] In Russia, rapid warming in Siberia fueled extreme fire seasons in eight of the ten years between 2000 and

2010.[27] China, Australia, and British Columbia have also seen tremendous tracts of forests burn since 2000.[28]

Bright spots for the planet's forests

Still, some bright spots exist for the planet's forests. Abandoned farmland is reverting to forest in parts of the eastern United States, Europe, and Russia, while regrowing forests in the eastern United States are believed to be benefiting from longer growing seasons associated with the changing climate.[29] In El Salvador and the Amazon, it has been reported that the abandonment of many farms is resulting in widespread secondary forest regrowth.[30] Evidence has been found that some forests are growing more vigorously in response to rising levels of carbon dioxide, although this effect is probably modest.[31] Community, government, and international efforts have had some success in slowing the destruction of tropical forests around the world—more about this in Chapter 12.

Worldwide estimates vary, but many experts believe that for now, forest die-offs are being offset by tree planting and the regrowth of forests in places where agriculture has been abandoned.[32] So far, analyses suggest that Earth's forests continue absorbing about a quarter of the carbon we humans are emitting.[33] But the overall trends of forest die-offs worry many of the world's leading forest experts, and tremendous uncertainty still exists about how continued warming will affect the world's forests.[34] One 2015 article by three leading scientists argues that we may be dramatically underestimating the vulnerability of Earth's forests to widespread die-offs under the coming hotter, drier conditions.[35] Yet we really have no idea whether forest die-offs will worsen over the next few decades, and if so, by how much.

This uncertainty is partly because we can't predict exactly how fast the climate will warm. But it's also because, quoting journalist Jim Robbins again, while they are a near miracle, "trees and forests are poorly understood on almost all levels."[36]

5 Agriculture

It's surprising that in all my geographical and professional wanderings, I have never seriously considered becoming a farmer. I love eating fresh-picked produce, working outside, and gardening. I spent a year after college working on an organic farm. I've visited many farms that are incredibly beautiful, where I'd absolutely love to live. I've helped and interviewed farmers, written about farming issues, and been to farming conferences. I'm always starting conversations with vendors at the farmers' market. But as far back as I can remember, I've understood that farming is hard, risky, relentless work; it takes rare talent, knowledge, and stamina to farm well. I admire farmers, but I don't envy them.

When we moved to the Willamette Valley, we found ourselves in one of the most fertile, productive, and diverse farming regions in the country—a place where many farmers actually make a decent living. Western Oregon, so far, has not experienced many of the extreme heat waves, droughts, and weather disasters that much of the United States and many other regions have recently suffered, and the regional climate predictions for the Pacific Northwest are not that bad. Adapting to the shifting climate may be a realistic enterprise in the Willamette Valley, at least for a while. However, in the places where the majority of the world's food is grown, climate change does pose huge challenges for farming—which is, after all, dependent on the weather.

Threats to agriculture from the changing climate

No human activity is more reliant on a stable climate than farming. Yet the growing signs that climate change threatens our agricultural systems have come as a bit of a shock. In 2007, the IPCC predicted that warming might benefit crop and pasture yields in temperate regions, offsetting any losses in the tropics. Those predictions now appear to have been wrong.

Agriculture experts have begun detecting some disturbing trends. First, there is the declining production of major food crops in most of the world linked with rising temperatures.[1] Then there are the many weather disasters since 2000 that have led to failed harvests in regions around the world.[2] Global food prices have been rising steadily since 2000, which, in combination with recent price spikes,[3] has worsened suffering for the world's poor.

The earlier optimism about the impact of climate change on food production came from several miscalculations. The first is related to the fact that plants take carbon dioxide from the air and convert it to sugars in order to grow. Since human activities are raising carbon dioxide concentrations, it was predicted that food crops would respond by sucking up some of the extra carbon dioxide and growing faster. Some early studies conducted in greenhouses supported this expectation, which led to optimism that crops would be "fertilized" by rising levels of carbon dioxide. Evidence now suggests that higher carbon dioxide levels don't boost plant growth very much under natural conditions and don't offset climate change's more harmful effects on plant growth such as worsening droughts, floods, and heat waves.[4]

Another prediction that isn't holding up is that although global warming might harm farming in the tropics, temperate regions would benefit from longer growing seasons and less cold. Models failed to take into account what a single extreme weather event or the spread of a pest can do to a crop. Indeed, increasing weather extremes are causing more crop failures overall.[5]

One study compared crop yields in the United States with detailed weather data across the entire growing season. Yields went up modestly with temperature until they reached a threshold: 29 degrees C (84 degrees F) for corn, 30 degrees C (86 degrees F) for soybeans, and 32 degrees C (90 degrees F) for cotton. Above these thresholds, yields dropped sharply. Using IPCC climate projections, the authors projected that under the slowest climate change scenario, yields of corn, soybeans, and cotton in the United States would decrease by 30 to 46 percent by the end of this century. Under the fastest warming scenario, yields were projected to fall by between 63 and 82 percent. The implications of these projections for global food supply are grim: corn and soybeans are two of the four largest calorie sources worldwide, and the United States produces 41 percent of the world's corn and 38 percent of the world's soybeans.[6] As it is, global consumption of rice, wheat, corn, and soybeans has outstripped production many of the years since 2000.[7]

Along with decreased predictability of seasonal events and worsening extreme events causing crop failures, rising temperatures expand the range of many agricultural pests and increase their ability to survive the winter.[8] Farmable areas and the length of the growing season will likely expand in places like Russia and Canada, but it won't be enough to offset the losses in places like Brazil, Africa, Asia, and the United States, and any newly farmable areas will be subject to the same destabilizing climate as the rest of the planet.

Recent scientific assessments acknowledge with increasing urgency that climate change poses significant and worsening threats to agriculture and food security.[9] A 2014 NASA report warns that drought is a growing threat to agriculture, as the aquifers underneath the world's most productive farming regions are rapidly being depleted.[10] The most recent IPCC report concludes that climate change has already negatively affected crop yields and predicts that climate change will continue to reduce food production, posing a serious threat to global food security.[11]

The size of this problem is unknown due to uncertainty about future greenhouse gas emissions, climate sensitivity, agricultural adaptation, population growth, and other factors that will affect our ability to grow food around the world. Since the green revolution, hunger experts have maintained that the world grows enough food to feed everyone adequately and that hunger persists largely because of poverty, waste, and distribution issues. But climate change and human population growth may spell the end of the era of global food surpluses. We have already put nearly all of the planet's potentially good farmland into production. As climate change increasingly destabilizes the weather, farming becomes riskier and more difficult, and crop losses become more likely.

The likely future: Rising food prices and hunger

The future holds its secrets, but I believe we have reason to worry about the possibility of massive starvation within the next several decades. Heat extremes, droughts, flooding, and ecologically novel pests and diseases all pose risks that are difficult to assess and presage events that are impossible to predict. Agriculture could go into steep, unanticipated declines in some parts of the world. If this happens, food prices will spike and the poorest people on Earth will be the first to starve. Africa, already home to nearly two hundred million malnourished people, is expected to suffer the biggest warming-induced cropland losses of any continent.[12] South Asia, southern Africa, parts of China, and Latin America are already challenged with large populations living in poverty and food insecurity, and are vulnerable to large potential crop losses with global warming.[13]

There are renewed worries about rising hunger rates in countries around the world, ranging from Uzbekistan and Mexico to Yemen and Haiti. Following forty years of falling food prices, global food prices rose at an average annual rate of 6.5 percent between 2000 and 2012, while price volatility has increased dramatically since 2006. Although many factors influence food prices, climate destabilization has been an important contributor to the rise in global food prices and volatility starting in 2000.[14]

As the scale of this challenge becomes clear, many agricultural experts have begun proposing adaptation projects: plant breeding to develop heat- and drought-tolerant varieties; changing planting times, irrigation, and residue management; changing crops; expanding crop areas into higher latitudes; and expanding irrigation systems.[15] However, agriculture is not just a victim of the changing climate; it is also a major contributor.

Agriculture's contribution to climate change

No matter how you measure it, the way we grow, process, distribute, and store food emits a lot of greenhouse gases. Depending on assumptions made and activities included, most estimates are that food systems are responsible for between 19 and 35 percent of human greenhouse gas emissions.[16] To avoid massive human suffering, the world must somehow learn to grow adequate food in a warming world while at the same time steeply reducing the greenhouse gases emitted by growing that food. This will require a worldwide shift to farming practices that are more resilient in the face of climate change, more productive, and more climate friendly.[17] No clear path exists to get there from here.

The issues with reducing agricultural emissions are so varied and so tangled with other human and ecological factors that they are difficult to summarize. Discussing the problem of livestock illustrates the complexity of the situation. The Food and Agriculture Organization (FAO) estimates the livestock sector alone accounts for 14.5 percent of all global greenhouse gas emissions, about half of all agricultural emissions worldwide.[18] Other estimates suggest the figure may be as high as 20 percent.[19] These emissions come mostly from (1) growing and processing livestock feed, (2) methane from ruminant digestion, and (3) manure.

Some experts advocate moving livestock from feedlots to rotational grazing on existing pasture to eliminate the emissions from producing livestock feed.[20] Well-managed pastures sequester carbon and offset emissions, but the amount of carbon sequestration and the feasibility of scaling up climate-friendly grassland management is hotly debated.[21] Other experts propose moving livestock off pastures into feedlots to spare grasslands and prevent deforestation, although this approach has been criticized as cruel to animals and highly polluting.[22] These two remedies are in exact opposition to each other.

Installing biogas digesters, improving manure management, and adopting other best practices could reduce livestock emissions,[23] but it's very hard to imagine how we could implement these proposals worldwide. One point of agreement is that animal agriculture is an inefficient and highly climate-damaging type of food production, and that global meat and dairy consumption, which has been rising steadily since the 1960s, must shrink if emissions from livestock are to be reduced.[24]

As our understanding of the complexity of the problem has grown, many farmers and experts have been experimenting with a range of solutions to reduce agriculture's contribution to climate change.[25] Nitrous oxide emissions,[26] which contribute an estimated 8 percent to greenhouse gas emissions worldwide, can be cut dramatically by reducing fertilizer use or shifting to organic farming methods.[27] Carbon storage in the soil can be increased by implementing conservation practices such as minimizing tillage, reducing erosion, reducing fertilizer use, composting, and organic farming.[28] Other ecological farming practices that return carbon to the soil include no till or perennial farming, reduced and more targeted irrigation, and agroforestry, in which trees are interplanted with other crops.[29] To be effective, these solutions must be tailored to individual regions, cultures, countries, economies, and crops.

Typically, farming methods that reduce nitrous oxide emissions or increase soil carbon storage have many other environmental benefits[30] but also require more tailoring to each location, smaller farms, more use of crop rotations, more labor, and fewer chemical inputs. Converting to climate-friendly agriculture and reducing animal farming, especially in feedlots, would require a massive transformation of farming on a global scale. No agreement exists about what this would take, if it is possible, or if such a conversion could produce enough affordable food to feed the world's growing population.

Meeting food needs in the new world

Some people believe we can solve the problem of global food security through technological advances. Bill Gates has been on a mission to end world hunger and thinks he can do it by funding and promoting the development of genetically modified (GM) crops that can produce more food and deal with climate change. Gates has reportedly given $27 million to Monsanto for the development of such GM crops and hopes to influence the direction of African farming.[31]

If you are a techno-optimist, as Gates is, you may believe that new genetic technologies and the intensification of industrial farming methods will allow us to produce more food for an ever-rising population even as the climate warms. My observation is that such technology-driven solutions tend to have unintended and often highly harmful side effects while underdelivering on miracles. Gates has been widely criticized for his philanthropy in this arena.[32] Also, as a biologist, I have a hard time imagining how inserting genes will allow plants to survive weather chaos. The one thing I agree with Gates on is that climate change poses an extreme and ever-growing threat to our ability to grow enough food for a rising world population.

While climate change threatens agriculture globally, individual locations and farmers will face varied fortunes as climate change proceeds. In the Willamette Valley and other pockets in the United States, some farmers and consumers are reestablishing local food systems as a hedge against the threats they fear from climate change, soil depletion, and water scarcity. Such diversified smaller-scale food systems allow for more experimentation and adaptation. Near my home, two new warm-weather crops—kiwi fruit and olives—have recently been planted, and winemakers are planting new grape varieties as the region warms. But at this point, these local food systems—even where they are thriving—make up only a tiny portion of people's diets.

Even as we begin to recognize the need to adapt farming to the changing climate and reduce the greenhouse gas emissions from agriculture, the challenges of meeting our food needs in this new world are probably greater than anything else we've faced in the modern era. As Bill McKibben has observed, the development and expansion of human agriculture has all occurred during a ten-thousand-year period of global climatic stability that perhaps could have gone on much longer.[33] This means that all modern agriculture is based on a relatively stable and predictable climate. Because we have disturbed this stability, we will no longer be able to grow food the same way we have been during the climate "sweet spot" we are in the process of disrupting.

6 Paradoxes

My dad grew up in Brooklyn, the only child of Russian Jewish immigrants. During my California childhood, we would travel east every year or two to visit his parents, often at Thanksgiving. My grandparents' tiny apartment couldn't hold us all, so we'd stay in Manhattan and travel back and forth to visit them by subway or taxi. I have lots of memories from those trips: the roar and shake of the well-worn subway cars going down under the East River, chilly evenings walking down Fifth Avenue looking at the Christmas displays, climbing the endless stairs to the top of the Statue of Liberty with my grandmother, and longing to join the ice skaters at Rockefeller Center. When my daughter, Lia, was just a toddler, I took her on the long journey to meet her great grandma shortly before she passed away at age 103. Then again in April 2011, when Lia was eleven, my mom and I took her back to New York, wanting to connect her to her history and share what we loved about the city.

We stayed on West 57th Street, next to Carnegie Hall and within walking distance of food, entertainment, art, and sights to keep us occupied endlessly. We had five days, and we had a ball. I have never been a city person, but on that trip the creativity of New York's artists and musicians and chefs and architects impressed me in a new way. In fact, I was both delighted and overwhelmed with the sheer possibilities of New York. I don't believe that the city itself had fundamentally changed since my visits growing up, but the world, and my perception of it, surely had.

We'd been living in Corvallis, Oregon, a place city dwellers consider a backwater, for more than a decade. The contrast with New York was stunning: in the city, the twenty-four-hour activity, the sights, sounds, smells, and tastes were unlike anything Lia had ever seen before. But it wasn't just the overstimulation that astounded me; it was the depth and breadth of the options—art, music, theater, food, fashion, architecture, and amusement to satisfy any taste, any time of the day or night. The city seemed to me bursting with endless, mind-blowing choices.

On that visit to Manhattan, I finally understood—for the first time—how it is possible to feel completely removed from the reality of climate change. When you are surrounded by everything from Tiffany's to FAO Schwartz, from extravagant productions of *Spider-Man, Wicked,* and *The Lion King* to lavish Oprah parties, when you see entire shops filled with Swiss chocolate, designer dog clothes, and silver baby slippers, you ask, how could we possibly be hitting the redlines of our planet? I experienced on a gut level how difficult it might be to believe that we are on the brink of a major crumbling of our life support systems.

We took that trip a year before Hurricane Sandy came ashore. We stayed on the *exact* block of West 57th Street that was evacuated when Sandy's high winds broke a crane on a high rise and the boom dangled precariously over the street for six days. As remote as climate change felt on our trip, of course it wasn't; its impacts were simply hidden from view—until they weren't.

The environmentalists' paradox

A paradox can be defined as a proposition or situation that seems to present a contradiction but in fact is or may be true. Paradoxes often feel both compelling and troubling, I think because as humans we are always trying to fashion a coherent story from all the information we take in. A paradox is the opposite of a coherent story. For me, the age of climate change is full of paradoxes, and visiting New York City before Hurricane Sandy illustrated one of them perfectly. I learned this particular contradiction actually has a name; it's called the environmentalist's paradox.

Here it is: we are in the midst of creating the worst human-caused environmental crisis ever, yet we live—more than seven billion of us—using ever more resources and with a rising average standard of living. As we degrade the planet and alter the climate, as human population continues to climb, we consume more, pollute more, and waste more—and nothing seems to be stopping us. My trip to New York brought this paradox into focus. Then I read an essay that named it and made it explicit.

The essay by Eric Wagner, titled "It's the End of the World As We Know It ... and I Feel Fine,"[1] argues that starting with Paul Ehrlich in 1968, environmentalists have been predicting an age of scarcity and famine, but it hasn't yet happened. "Why," the author asks, "has the human condition—incontrovertibly and inconveniently, it would seem—gotten better and better, even as most ecosystem services have declined?" This paradox has taken on a dramatic new turn with the advent of climate change. Wagner presents evidence from a 2010 analysis published in *BioScience* that the Human Development Index, a widely used measure of life expectancy, literacy, and income, has increased in all major regions of the world over the past thirty-five years, while the global poverty rate has dropped by nearly 75 percent, in spite of widely documented ecological declines. The most recent Human Development Index information I could find, from 2014, show HDI values have continued to improve slightly.

Wagner offers several possible explanations for the environmentalist's paradox but comes up with no conclusive answers. Perhaps the damage we are doing just hasn't caught up with us yet but soon will—but this explanation is untestable at the moment. We are left with the question: if what we are doing to the planet is so bad, why are we able to squeeze out of it more and more food and resources to fuel growth and consumption?

Until my 2011 trip to New York, I hadn't perceived this as a paradox so much as a lack of discernment—we are poor judges of where the actual redlines are and how exactly we'll come up against Earth's final limits. Warning signs are everywhere, but we can't see the location of the actual precipice. But that spring trip to New York helped me understand that the environmentalist's paradox is real, and it helped me see how people might truly not believe in the threat posed by climate change. Most Americans today live a largely indoor life, buffered from worsening climate extremes and crop losses. Gazing from the observation tower of the Empire State Building at the world below, eating some of the best Italian food I've ever had, strolling through Columbus Circle at night with fountains spouting and city lights sparkling, I *almost* couldn't believe we're pushing the planet's climate toward a major and possibly irreversible shift. I finally *got* our collective failure to feel in our bones that climate change is utterly real and to act accordingly.

The paradox of scientific certainty and public disbelief

Contemplating the environmentalist's paradox in New York made it a little easier for me to understand a second paradox—that even though the science of climate change and its causes is rock solid, many people deny or question its nearly universal acceptance among climate scientists. Here is one of my favorite statements about this, from author Paul Gilding: "If you cut down more trees than you grow, you run out of trees. If you put additional nitrogen into a water system, you change the type and quantity of life that water can support. *If you thicken the Earth's carbon dioxide blanket, the Earth gets warmer* [italics mine]. If you do all these and many more things at once, you change the way the whole system of planet Earth behaves, with social, economic and life-support impacts. This is not speculation, this is high school science."[2]

By including Gilding's words, I don't mean to imply that climate research amounts to high school science. Not at all. But the underlying processes at work are widely understood and fully accepted in the world of science. More than 97 percent of climate scientists, based on the evidence, are convinced that human-caused climate change is happening.[3] After the IPCC's fifth draft report was released, several leading climate scientists were asked to explain the level of scientific certainty that man-made climate change is under way. The scientists said that the 95 percent probability established about man-made climate change is similar to the level of certainty we have that cigarettes are deadly.[4]

In spite of this, many people hold tight to the notion that the science is controversial. A 2013 survey found that 43 percent of Americans believed that climate scientists had yet to come to a consensus about the cause of climate change.[5] In a 2015 survey, 37 percent of those questioned said scientists do not agree that Earth is getting warmer due to human activity.[6]

Trying to come to terms with this paradox—that the science is settled among scientists, yet the existence of scientific consensus is unknown to a significant portion of the public—has caused me much frustration. It often seems like willful ignorance or self-destructive magical thinking. And I know this misunderstanding has been fueled and exacerbated by a well-financed disinformation campaign and lots of poor journalism (more on this in Chapter 7). But sometimes I think the truth has just been too hard for many people to face. Too overwhelming, too scary, too much to comprehend and deal with.

The paradox of twin possibilities

Which leads me to the last paradox on my mind, in some ways the biggest one of all. Climate change, and all that it may bring, provides the twin possibilities of both the greatest tragedy we have ever brought down on ourselves and other living things, and the biggest opportunity humanity has ever had to alter the path of human civilization, so often dominated by greed, destruction, domination, and exploitation.

Looked at purely from the standpoint of the data, huge, destructive changes will be caused by rapid temperature rise, ocean acidification, melting of Arctic ice, sea level rise, a sixth global mass extinction event, increasing climate instability with more severe and destructive weather, crop failures, and threats to our life support systems. I can't see a way we're going to navigate climate change without widespread human suffering, many losses of other life forms, major political instability, and the likelihood of increased violence in many parts of the world. Perhaps the global food system will crumble. Perhaps the world's fisheries will undergo massive disruptions or even collapse. Probably seawater will inundate human population centers. The Amazon jungle may begin to die. Coral reefs will very likely disappear.

These projected changes, if they are not contained and reversed very soon, are expected to threaten human well-being and potentially even human survival. We are bringing disaster upon ourselves, our children, and other living things on a scale that is almost impossible to grasp.

And yet ... how can I say this? Climate change may also offer us an opportunity to change ourselves and our culture. You may understand what I'm getting at if you have ever, even fleetingly, felt that human civilization is too often a vehicle for exploitation of the vulnerable and the voiceless; or if you've ever wondered while studying history if we really learn anything from our mistakes; or if you've ever felt despair when looking at a polluted river, a desperate shantytown, an ugly new shopping mall in a previously wild meadow, or really any indicator of human folly. If you have ever thought that taking over so much of Earth and leaving so little isn't right, this paradox might resonate with you.

We humans are often a greedy, selfish, insatiable species. I've been heartbroken so many times walking through clear-cuts, poisoned fields, bulldozed meadows, devastated mountain mines, trashed gas fields. Rarely does nature get any kind of fair deal. Seen in a certain light, climate change may allow us—force us—to reconsider all this. Is it possible that as the signs of major climate disruptions continue to strengthen, we will find ways to drastically reduce emissions to prevent the worst projections from coming true? Is it possible we will rein ourselves in and help each other endure the changing weather?

Sometimes people are able to unite in times of crisis in ways we never seem capable of otherwise. All evidence points to disaster, yet perhaps we'll be our best when adversity forces us into a corner. There are those who believe we have this potential. This may be the greatest paradox of all: all the facts point toward tragedy, yet the potential for human ingenuity and goodness in a true catastrophe is out there as a tantalizing possibility.

Part 2 | The Question of How to Respond

David Roberts, a wry and witty blogger at *Grist,* gave a TEDx talk in 2012 titled "Climate Change Is Simple" that you can watch on YouTube. In the talk, Roberts makes three simple points: our present course leads to unthinkable climate catastrophe; to stabilize temperature, global greenhouse gas emissions need to peak by 2017 to 2022 and then decline rapidly every year thereafter; given current politics, reversing global emissions that rapidly seems impossible. His concluding sentence is, "That is where we are, stuck between the impossible and the unthinkable."

I agree with Roberts's assessment. I've just taken you on a brief tour of what we know—and don't know—about climate change and what is looking like a rapidly approaching global catastrophe for humans and other living things. And once you've truly absorbed the news, you've got to ask this one question: what shall I do about it? As I've been searching for a way to think, feel, live, and act, everything keeps coming back to this paradox: we each need to do much more, and do it fast, yet as long as so few of us are fighting for action, every action each of us takes is hopelessly small and inadequate. In short, getting there looks impossible.

So in choosing how to respond to climate change given our current predicament, I definitely asked myself whether it might be valid to understand the facts and their implications, admit to being powerless, and simply live your life as best you can. It can seem reasonable, even rational, but is it a justifiable reaction? And the answer I kept coming back to, over and over, is no, it's really not okay. If you follow that thread even a short distance, you come upon results that are both unacceptable and unconscionable.

You might care most about a different loss or disaster from what I care most about. Is it the disappearance of your children's chances for a safe and healthy future? Is it that millions of species are threatened by the rapid climate destabilization we are bringing on? Is it the potential loss of the comforts and safety of modern society—transportation, hospitals, affordable food? Is it that entire island nations face the disappearance of their homelands? Is it that some of the world's poorest people are already facing more hunger because of climate change and will suffer the most? Is it that governments may collapse? Is it the possibility of terrible wars and carnage over water, food, or land as the climate destabilizes? If you've read to here, I imagine you understand that these are not distant possibilities for the next century but threats we face in our immediate future. I believe that if you really look unflinchingly at whatever is most unacceptable to you, you can't simply turn away. So that means facing our situation and trying to do something about it.

In the chapters that follow, I take you along as I look for both emotional and practical responses to global warming—how to feel and how to act. The feelings I've felt while accepting the full implications of climate change are mine, but people I know tell me these feelings are widely and deeply felt by many others. I also examine things to do, including what I believe are some of the most promising actions people are taking to address the climate crisis. Last, I'll tell you about my own choices, and where writing this book has taken me.

7 Blame and Moving On

Sometimes I just want someone to blame for the mess we're in. We've known for *decades* that our actions are altering Earth's climate. Since 1979, when the National Academy of Sciences first convened a panel to examine the evidence for human-caused global warming, scientific understanding has grown stronger and more certain year by year. The 2013 IPCC report, a consensus document on the state of the knowledge, concluded with 95 to 100 percent certainty that human activity is altering the climate.[1] This is, scientifically speaking, as good as it gets. Yet over these same decades we've failed repeatedly to alter the trajectory of rising greenhouse gas emissions.[2]

For me, our failure to make progress during the first term of the Obama presidency was especially heartbreaking—after hearing Obama say in his nomination victory speech, "We will be able to look back and tell our children that this was the moment when … the rise of the oceans began to slow and our planet began to heal."[3] The global economic crisis hit just before he took office, and President Obama became silent about climate action for many years.

As climate scientists became ever more urgent and united in their warnings, as climate destabilization advanced, our inability to halt rising emissions started to seem absolute and irreversible. I felt more and more that our failure to safeguard the future of our children and grandchildren, other species, and our livable climate must be someone's fault.

Who is to blame?

Environmental author Bill McKibben has been writing about climate change since 1989 and today is one of the world's leading climate activists. A few years back, McKibben wrote about who is responsible for our failure to take action. In one essay, he blamed Congress, for being too lazy and weak-minded to tackle something as complex and contentious as climate change.[4] In other writings, he blamed the unchecked power and influence of multinational corporations over the U.S. government. Especially, he blamed the fossil fuel industry and its allies for their well-financed disinformation campaigns to discredit and attack climate scientists.[5]

In fact, entire books have been written about climate change disinformation efforts and their chilling effects on progress in addressing climate change. For two examples, take a look at *Climate Cover-Up: The Crusade to Deny Global Warming* by James Hoggan with Richard Littlemore, and *Merchants of Doubt* by Naomi Oreskes and Erik Conway. Exxon leaders were reportedly told by company scientists in 1977 that emissions from use of fossil fuels would warm the climate and could endanger humanity. They spent the next three decades studying the problem while publicly denying the threat and spending money to mislead the public about the facts of climate change.[6] Additional villains in the war against facts include the Koch brothers, industrial billionaires who as of 2013 had reportedly spent more than $79 million on front groups promoting climate change denial claims.[7]

Another target of blame has been climate scientists, for failing to translate their work effectively for nonscientists, to defend it adequately, or to make policy recommendations. While I have sometimes chafed at the circumspect and cautious language used by climate scientists, I can understand and sympathize with them. These scientists, particularly in the United States, have faced personal and professional harassment, character assassination, and even death threats for simply carrying out and publishing their research.[8] Sadly, efforts to intimidate climate researchers—even by Republican members of Congress—continue to this day.[9] If many are especially cautious in communicating about their work, this is a reasonable reaction to being attacked and threatened for simply doing their research and explaining their findings.

Others have blamed the media for the public's failure to take action, and indeed, news coverage of climate change was mostly misleading, inadequate, and at times flat wrong until coverage began to improve in 2012.[10] When I began writing about climate change in 2006, I was astonished that essentially all the coverage provided an "opposing scientific viewpoint" questioning or rejecting the evidence that the climate is changing or that human activity is responsible. We now know that those providing this opposing scientific viewpoint were being paid by corporations to sow confusion and doubt.[11] However, at the time journalists typically gave their views equal weight, creating the impression that climate science was wildly uncertain even as the scientific consensus had been largely established. (One of the most widely quoted denier climate scientists, Dr. Wei-Hock Soon, took $1.2 million over the course of a decade from the fossil-fuel industry in exchange for producing papers and congressional testimony that greenhouse gases pose little risk to humanity.)[12]

This practice of offering false balance is finally beginning to give, especially in the print media. However, cable news shows continue to ignore or distort the facts about climate science and even offer misleading "debates" on settled science.[13] I believe the damage that has been done is tremendous and continues to this day. Polls suggest that as late as 2015, only about half of Americans understood that human activity is the primary cause of global warming.[14]

Some blame our failure to act on those who speak too bluntly about the dire threats posed by climate change. This argument is based on the contention that too much bad news shuts people down and discourages them from taking action. For example, an article in *Science* magazine argued that mountains of grim climate news were causing "climate fatigue" and that "a drumbeat of dire warnings may be helping to erode U.S. public concerns about global warming."[15]

A new variation on this theme suggests that negatively framed media coverage is turning people into climate cynics. Two studies reported on in 2015 found that news coverage most often portrays climate action as costly and unsuccessful instead of manageable and effective. This negative framing influences the public to feel despair, alienation, and cynicism.[16] The implication here is that the media is influencing our failure to act, and the media should spin climate stories more positively to help spur us to action. I agree completely that what we read, listen to, or watch on TV influences how we feel, what we think is possible, and how we act. What bothers me about these analyses is the idea that solving the problem is as simple as fixing the negative framing of the stories we are told. If we are so numb and unquestioning about the information we are being given, and if the media is generally only reporting costly and failed climate action, aren't those the real issues?

Finally, we can always blame ourselves. Few of us, myself included, have made taking action to fight climate change the top priority it deserves to be. Few of us are doing anything at all. Our failure is constantly reinforced by living in a fossil-fuel-powered civilization; I like indoor heating and taking my car when the weather turns damp and cold. It's hard to make time and find money to insulate my house and get solar panels. It's also hard to say no, consistently and continually, to buying cheap stuff I don't really need, even if making that stuff is changing the climate. Working for societal change—writing elected officials, speaking out, voting, organizing to fight climate change—is even harder for most of us. So we alternate between blaming ourselves for not doing more and listing all the reasons doing more is too difficult or pointless.

Indeed, a new and growing body of research examines why we haven't taken action to fight climate change when the evidence is so overwhelming and the threats are so massive. George Marshall, Kari Norgaard, and Per Espen Stoknes have all written books about why the public has failed to act.[17] Marshall describes climate change as a "wicked" problem because it is contradictory, constantly changing, slow moving, and without an obvious enemy, without simple causes or solutions. Norgaard describes the phenomenon of "socially organized denial," in which people know about climate change and see its impacts yet at the same time find it unimaginable and remain silent about it. Stoknes discusses what he calls the five main psychological barriers to climate action—distance, doom, dissonance, denial, and identity. These experts all offer explanations that say that human nature combined with the nature of this threat are to blame for our failure to act.

Time to move on

But wait. In spite of all our failures, all the missed opportunities and cynical, even evil acts that have contributed to where we are today, I believe in my heart that it is better to move on than to look back searching for villains. I agree that holding some parties accountable, such as prosecuting Exxon for its past illegal and immoral behavior, is worthwhile.[18] I think it is worth acknowledging the steep odds and the human weaknesses we face. But I also believe it's time to let go of our past failures and assess where we are and where to go from here. It's time to move on.

Here's what I see right now. Many climate change deniers and obstructionists are rich and powerful, *but they are in the minority*. Even though a huge amount of money continues to flow from corporate sources to confuse the public and prevent meaningful action on climate,[19] we have some decent news coverage about climate change, without quotes from deniers in every story (with cable TV news as an exception).[20] Some major corporations are acknowledging the threats posed by climate change, setting goals for emissions reductions, planning for a price on carbon pollution, and even supporting regulations to bring down emissions.[21] Important military, religious, scientific, and political leaders are speaking out about the severity of the threat and the need for immediate action.[22] Recent actions in the United States such as finalizing the Clean Power Plan and rejecting the Keystone

XL oil pipeline open up the possibility for progress on reducing emissions and combatting climate change. In December 2015, the Paris climate accord signaled that essentially every nation on Earth finally recognizes the need to address climate change.

Getting a handle on Americans' beliefs on climate change is tricky; poll results fluctuate, affected by everything from poll wording and economic trends to the recent weather.[23] Concern about climate change fell sharply during the economic crisis beginning in 2009.[24] However, polls conducted at the end of 2011, immediately after the United States had experienced fourteen major weather disasters, found increasing American acceptance of the scientific consensus on climate change, partly driven by personal experience with extreme weather.[25] There is even some evidence that Pope Francis's call for climate action in the summer of 2015 influenced American public opinion.[26]

We can get some sense of public attitudes by looking at polls taken over time. Figure 4 shows responses to the question "Do you think global warming is happening?" asked repeatedly between November 2008 and March 2016 by the same pollsters. Results indicate that most Americans (generally between 60 and 70 percent) accept the scientific consensus that climate change is occurring. Some are uncertain, and a fairly small minority (generally between 10 and 20 percent) don't believe it's happening. Yet this poll also suggests many Americans continue to be confused about climate change's cause: only 53 percent of those surveyed in October 2015 think that climate change is due mostly to human activity.[27]

Another important finding from tracking polls is that a majority of Americans say they worry a fair amount or a great deal about climate change. This number, tracked by Gallup over the past fifteen years, has fluctuated

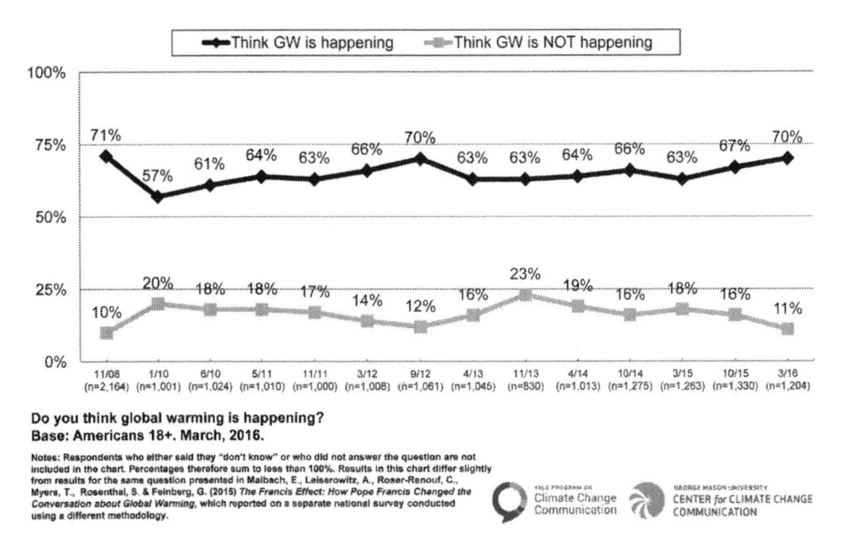

Figure 4. Global warming poll results, November 2008 to March 2016. Source: A. Leiserowitz et al., *Climate Change in the American Mind,* March 2016, Yale Project on Climate Change Communication and George Mason University Center for Climate Change Communication.

between 51 and 72 percent.[28] However, in spite of this level of worry, about three quarters of Americans say they rarely or never discuss global warming with family and friends.[29]

My overall assessment is that today most of us get that climate change is happening and that it is or will soon be a big problem. Some of us get why, but confusion persists about human activities being the primary cause of climate change. Many of us worry about global warming, but we don't like to talk about it. Many of us don't realize just how serious and imminent the problem is, or we don't know how to take meaningful action.

Even though climate change news coverage has improved over the past decade, climate change is still, in my opinion, under- and inadequately reported. Climate success stories still get almost no mainstream news coverage. Until now, the fossil fuel industry and its allies have exerted tremendous control over our federal government and have blocked many actions to shift us away from fossil fuels. There are some signs that the power of those opposed to taking climate action may be starting to erode—more on that later. But the future is unknown. This is the point we are at, and it is from here that we must move on.

8 Guilty Pleasures

I love my purple 4Runner—a 1998 six-cylinder manual, with a gray cloth interior, a sunroof, and more than 200,000 miles on the odometer. We name cars in my family, and her name is Jesse. I love driving her, the great visibility and high clearance, the well-timed shifting of gears, the deeply familiar shape of the driver's seat against my backside. I like the way she looks too, her simple hardworking lines, no frills but stylish. And then there are all the memories: road trips, camping trips, interstate moves, shuttling for river trips, navigating icy roads to go skiing, commuting pregnant to do fieldwork barely able to fit my belly behind the steering wheel.

There have been close calls too. One winter in central Oregon, I hit a patch of sheet ice on a cold, bright morning with my family in the 4Runner. We spun and slid off the road with what seemed like dizzying speed but stopped with a jolt upright and unhurt two inches from a hefty ponderosa pine tree. We painted a green four-leaf clover on that pine tree to remind us of our good luck. A decade later, we used Jesse to teach Lia how to drive a stick. Jesse has never broken down, and as she gets older and more scraped up, I only grow more attached.

While we're on the subject of confessions, I love to travel. I've been to the Grand Canyon; the Baja Peninsula; the Peruvian Amazon; the Atacama Desert; Costa Rican cloud forests, volcanos, and beaches; and, perhaps my favorite destination of all, the Galapagos Islands. I've been other places too—Alaska, Hawaii, Greece, Ecuador, Mexico, England, France, and Spain. Someday I hope to go to Africa.

When you're writing a book about responding to climate change, feeling guilty about owning an SUV and loving to travel seems reasonable. And I have. I've wrestled with how to live and what to give up.

Wrestling with how to live

I've never liked to shop, but I spent much of 2013 so sad and upset about the wastefulness and greenhouse gas emissions of unnecessary stuff that I truly hated shopping for anything other than food. Many times I'd set out on an errand to get something we "needed," decide at the store it wasn't necessary or I could do it another day, and come home empty-handed. I cut back on driving and wrestled with whether I could possibly sell Jesse, but I just couldn't do it. I listened closely to friends and colleagues discuss how they make their choices about participating in the fossil-fuel-based economy. Those who fully accept the reality and implications of climate change all seem to go through a stage of serious guilt. Many of us stay in this stage indefinitely.

But the grappling has gotten easier for me, and I'm approaching a workable resolution. I've realized that who you spend time with influences how you judge what's right and what's okay. I remember reading somewhere about a study that showed that if you have a circle of friends who are overweight or who smoke, you are much more likely to be overweight or smoke yourself, because unconsciously you see these things as normal and act accordingly. When I started writing this book, most people I knew were not terribly aware of or educated about climate change. I felt like a fanatic. As I got deeper in, I met climate change writers, activists, and scientists. Comparing myself to these people, I judged myself harshly—for my 4Runner, my plane travel, my frequent sojourns to central Oregon. In time, I saw that I could neither ignore the climate implications of my actions nor sacrifice the pleasures of my life in order to reduce my carbon footprint to its lowest possible level. Eventually, this allowed me to find some compromises that work for me.

It's a balance that I had to find, and it took time. Being one person, it can feel like it doesn't matter much how I live in the carbon-burning economy; but I had to weigh this against the ethical dilemma of participating (in however small a way) in a system that is destabilizing our climate. Really, it came down to deciding what felt worth doing. I sat down to summarize my list of personal choices and compromises. They may not be your answers, but listing them here may help you find the way to your own.

My choices and compromises

One change I've made is in making decisions about new purchases, I include some awareness that each choice has a tiny implication for Earth's carbon balance. For example, when our 80-gallon hot water heater was on its last legs, I researched replacement options, including tankless models that heat water only when you need it. I spoke with several skeptical plumbers who told me it wasn't worth the energy savings to convert it, or that tankless water heaters didn't work well. They tried to talk me out of it, but I became convinced it made no sense to keep 80 gallons of water hot all the time. Eventually I found a plumber who had experience with tankless water heaters and was willing to install one. It works fantastically well, we save energy every day, we got a good tax credit, and it wasn't hard or very expensive. When I do this kind of thing, I feel like I'm making a very small change in our culture by saying that to me, this matters.

I've developed a test when considering changes that could reduce my greenhouse gas emissions. For me, the test has become easier and easier to administer with practice: if it's affordable, brings no significant hardship, and is clearly a superior choice for the atmosphere, I make the change. If I can't stick to it, it probably doesn't pass the test.

For example, if I have errands, I do them all in one car trip if at all possible. If it's easily walkable, I walk. I find one day a week to do no driving. For some years, we paid a small monthly fee to our electricity company to subsidize renewable electricity production. This past fall, once the price had fallen enough to be within reach, and the efficiency had gone high enough that we could generate a good chunk of our own electricity, we installed rooftop solar panels on our house. We do meatless Mondays. I buy less stuff, and I try to buy things that will last. We choose lots of locally grown, organic, and in-season foods. I'm a total stickler about not wasting food.

When I'm considering travel, I think carefully about whether the trip will do some real good or enhance my family's happiness. If the answer is not a strong yes, I say no. If the answer is yes, I give myself permission to do it. Several years ago, my sister Karen invited us to go to the Galapagos Islands, and we went. It had been a lifelong dream of mine to see the Galapagos and the life there, and it was perhaps the best trip I've ever been on. Our lives were enriched: it was truly mind-expanding to explore a key birthplace of Darwin's evolutionary theory. Every time I think of it, I feel happy. Other times, I've turned down trips that sounded fun but just didn't feel important enough to burn the carbon to get there.

Moving past guilt

Once I began to make changes that reduced our family's carbon footprint and also accepted that my completely giving up things like driving and travel wouldn't stop the climate crisis, I was able to make an agreement with myself not to be consumed with guilt over my choices. One of the most destructive and cynical ideas spread by those fighting climate action is that if you use fossil fuels, you don't have the right to work or speak out against the corporations or systems promoting their continued dominance. I reject this argument; I strive to lower my impacts on the climate, I work for change, and I refuse to be flooded by guilt about my remaining carbon footprint. Still, I'm aware that within my lifetime, I will very likely witness the human world brought to its knees because collectively we are failing to adequately respond. I try to act in ways that ensure that if and when climate catastrophes strike, I will be able to look back on my choices and not regret them. And yes, I still own my purple 4Runner.

9 Grappling with Despair

I remember vividly a handful of times I've felt despair contemplating climate change—the bottomless kind, when you feel that you know the meaning of *heart-broken* for the first time. We are knocking Earth's climate system out of the stability it's been in, and the results are becoming dramatic, largely negative for life, and possibly irreversible. I know of no way to grasp this without feeling very bad about it.

Picturing a dark future

Once, I lay awake in the middle of the night after attending a talk by environmental law professor Mary Christina Wood. She was developing a legal framework to use the judiciary system to combat climate change.[1] Wood is an eloquent speaker and a mother of three; that evening she had spoken about our obligation to protect the atmosphere for future generations and how badly we are failing to do so. It was one of the first and clearest times that all of the extraordinary facts about climate change were fully laid out before me.

I've thought about that night often, because going to that lecture was one of those small decisions that ends up altering your life. Wood's presentation helped me perceive the full implications of destabilizing the climate and understand that failing to act is morally wrong. I am usually a sound sleeper, but after that talk, I lay awake in my bed feeling hopeless and alone. I believe that I had never fully realized before that night just how desperate conditions on the planet might become in Lia's lifetime. She was six at the time, an observant, extremely sensitive child who hated conflict and shrank from loud noises. Even the cafeteria at school was overwhelming to her, and she'd often come home hungry with her lunch uneaten. Imagining her in a disintegrating society torn apart by climate change filled me with misery. I saw the possibility more clearly than ever before of where we might be heading: a world made harsh, difficult, and dangerous by our own failure to change course in time. It was dizzying to have the weight of our species' behavior coming down on me in the darkness.

The end of natural beauty

Another time, I was writing about the wildfires that had been burning in central Oregon. Looking at the vast vistas of blackened forests as we traveled through the Cascades, I was shocked by how bad things looked, how little recovery I saw. So many fires—named for places they've destroyed: Fossil Creek, Cache Mountain, Black Crater, Rooster Rock, Shadow Lake, Pole Creek—have burned here since the new millennium began that the views are dominated by a patchwork of fire-scarred forests standing in varying states. For a time, whenever I'd pass through the central Oregon Cascades, I felt that hollow kind of dread that sits deep in your chest; it seemed to me then that the land would never green up again. I felt anguish that these gorgeous mountain forests that I had only recently begun to explore and to love were incrementally dying.

There is a feeling you can get—perhaps you know it—when it seems that all the beauty you treasure in the world is being diminished and inexorably snuffed out, bit by bit. I wanted to stop time, or really to go back in time, and hold on to the natural world as I had always known it. I felt certain it was all going to disappear, and I was bereft.

My failure as a parent

A third time I remember feeling despair was on a day when I felt like the ultimate failure as a parent. Lia has always been a keen listener, sometimes hearing quiet conversations from across the house. She was eight or nine at the time and playing in another room while I was discussing climate change with a friend at the kitchen table. She wandered in and asked, "Mom, will we be okay?" with a worried face and a plaintive note in her voice. "Yes," I replied, "we will be fine."

I think I babbled on to tell her that if we didn't stop climate change, other people less fortunate than we were, people in other parts of the world, and plants and animals would be harmed, so that was why we needed to work hard to fight climate change. But we would be safe. It felt like a lame and shady response. I felt awful scaring her about something she was powerless to change, and I also felt like a liar for reassuring her about something I was so profoundly afraid of myself, something I feared threatened her future. Worst of all, I felt sick that I could not protect her.

Our collective failure to really care

Finally, I've periodically felt despair about our collective failure to really care. This comes up for me talking to friends or acquaintances about climate change and watching them check out, or reading about a poll that finds that half of us say we aren't worried about climate change at all. Many climate activists I've talked to are angry and frustrated that corporate money is all but running our government, and I agree this is deeply upsetting. But I actually feel worse about the collective apathy of the American public.

Most of us have been living as if our climate isn't changing; our forests aren't dying and burning; our ice caps aren't melting; our oceans aren't warming, acidifying, and rising; other species aren't moving, adapting, or going extinct; and the weather isn't getting more extreme, destructive, and unpredictable. The reason I find this so heartbreaking is that I believe if a good-size chunk of the public were energized to take action, we could take back the government enough to pass a comprehensive climate bill that could dramatically reduce fossil fuel use and greenhouse gas emissions. Many people I know don't think this is possible, so I guess in a weird way this makes me an optimist.

Maybe it's possible to fully face climate change without feeling despair, but I don't think so. I believe that for most of us, this is an unavoidable stage. I think we all fear despair—I know I do. If you've been on this planet very long, you have probably experienced tragedies or losses that have brought you heartbreak, as I have. Remembering how it feels, I'm not anxious to return to this place, and I'm afraid of getting stuck. Yet feeling grief signifies that you get it in your gut—that climate change is personal because the things *you* love are in grave danger. I believe this understanding is necessary before you can fully respond.

The despair I felt in facing climate change could only pass once I acknowledged the validity of this response and found a way to take positive action. I came to understand this meant contributing to a larger movement, but for quite some time I didn't know what that might be.

10 Seeking Solace

In the fall of 2011, I was granted a writer's residency at the H. J. Andrews Experimental Forest in the western foothills of the Oregon Cascades. I remember that fall as a low point in my exploration of climate change and in my life. I was discouraged and grieving about all I was learning. I had temporarily lost my generally optimistic outlook, and although I was functioning, I wasn't feeling much pleasure in daily life. I love the Andrews Forest in the deep, comfortable way that I love some of my favorite childhood haunts, and I went there hoping for relief and clarity.

The Andrews and its immediate beauties

I arrived for the residency on a bright, cold Sunday in mid-October, planning to write but also hoping to find solace in nature. I don't read a lot of poetry, but Wendell Berry's "The Peace of Wild Things" had been on my mind a lot when I arrived.

When despair for the world grows in me
and I wake in the night at the least sound
in fear of what my life and my children's lives
may be,
I go and lie down where the wood drake
rests in his beauty on the water, and the great
heron feeds.
I come into the peace of wild things
who do not tax their lives with forethought
of grief. I come into the presence of still water.
And I feel above me the day-blind stars
waiting for their light. For a time
I rest in the grace of the world, and am free.

I was housed at the research station, a collection of bare-bones apartments, research labs, offices, and a cafeteria surrounded by forest. After checking in, I began exploring Lookout Creek and the nearby trails until the light began to fade. After an early dinner and some restless reading, I wrapped myself in my sleeping bag and went into a fitful sleep, missing my family. I struggled the next morning to settle in with the expanse of unstructured alone time that lay ahead. Then I headed out into the forest to explore, taking a notebook, snacks, and water.

The Andrews is not a wilderness area. It is set aside as a research forest; during my visit, studies were under way examining logging methods, spotted owl demographics, water cycle dynamics, bird and mammal diversity, and air quality, among other things. I saw many signs of experiments as I explored: research plots, treatment markers, flagging tape, and even occasional field crews. During my residency, I visited labs, talked to scientists, and learned about their research. Because I used to do and love ecological fieldwork, I thought I'd be fascinated by the science being done at the Andrews. And it was kind of cool. Still, it was the exuberance of nature that got to me right from the start.

What stays with me most are memories of the immediate beauties of the thriving forest: the owl calling at the decomposition site, the give of the moist spongy ground under my feet in the filtered forest light, the American dipper diving under the icy clear water in Lookout Creek and coming up dry, the huge dead trees lying across the gravel riverbed haphazardly like so many giant dropped matchsticks, the seemingly unlimited shades and textures of greens and browns sparkling among the moving shadows. It was all lovely to me.

Right away, being at the Andrews began washing away the chatter in my brain, drawing me out of my worries and into the forest, allowing me to think and feel and write. I began to feel better. By my second day, I was feeling moments of pleasure and beginning to release the despair that had been haunting me. Over the course of the week, as I hiked and wrote and spent time in the forest, I regained some perspective and remembered some long-forgotten knowledge.

Nature as my church

The first thing I remembered was that I find comfort and peace in places belonging mostly to other creatures, places not dominated by humans. Being alone at the Andrews gave me a grounded feeling that the world was as it should be and that I was one with the world. Being in the natural world is my spiritual practice. So many of my best moments have happened when I'm doing without the protection and safety of civilization: no grocery store, heating, air conditioning, phone, microwave, no inside shelter. This reminds me of something Joseph Campbell, the great mythology expert and teacher, once said: "I don't believe people are looking for the meaning of life as much as they are looking for the experience of being alive."[1] Being outside needing to fend for myself in a magnificent land, I am most likely to experience being alive.

I found that the thought that runaway climate change may be where we're heading didn't hurt me as much at the Andrews. Hiking the trails, sitting on the riverbanks, climbing the ridges, I felt the vitality, resilience, and aliveness of nature. I didn't care about Bill McKibben's eloquent arguments in *The Death of Nature* (his first book about climate change) or the overwhelming data showing that we have already altered the atmosphere to such an extent that every living thing will be affected and many unique life forms will be threatened with extinction. Life in the forest was beautiful, and it was thriving. As much as the future is uncertain, being in living nature was an antidote with the power to transport me beyond my worries for a time.

The power of good boundaries

I love maps, and as I explored the forest I pored over maps too. I admired the clearly defined and well-chosen boundaries of the Andrews. The forest covers the entire 16,000-acre watershed of Lookout Creek, and that's it. It's one of the real beauties of this forest; all I care about, you could say, is within these boundaries; all I want to learn and know is right here. There is great comfort and containment in that. The boundaries of the Andrews allow the scientists here to do their work unhindered by what's outside, and on a manageable scale. Yet the Andrews Forest is also a microcosm of the mountainous, wet, westside forests of Oregon's Cascades. Lessons learned here can be applied elsewhere, though not everywhere. I remembered during my time in the forest that I had to work on building good boundaries too.

Better boundaries will allow me to put aside the constant awareness of all I cannot hope to influence and all that might happen in the future. I need stronger boundaries around me, around my family and friends, around my community, the nearby lands I love, and around my present day, week, month, and year. I still have a lot of work to do on building good boundaries, but my time in the forest reminded me what they look like.

The mystery of other living things

At the Andrews, the beauty all around me woven from layers of living creatures and exquisite land gave me a longing to understand more about what other beings know, how they think and feel. I thought again about Wendell Berry's poem: we believe we alone are a species who worries, anticipates, plans, and regrets. We believe that we suffer and the wood drake doesn't because we think but birds just live. One day, when I thought I heard a northern spotted owl calling, I found myself wondering if spotted owls have any idea that while we humans have destroyed most of their habitat and cut down their forests to build stuff, we have also spent millions of dollars and countless hours studying how they live and trying to prevent their extinction. Do they wonder why we put mice out to bait them in, catch them, band them, weigh them, and then let them go?

Many studies suggest that plants communicate with each other. Scientists have found that walnut trees emit aspirin-like compounds when they are under stress. One human interpretation of this is that they may be warning other trees of impending problems. Other research on plant defenses suggests that plants release chemicals that may act as warnings, invitations, and possibly even pleas for help.[2] Is this evidence of plant consciousness?

There may be no way for our species to bridge the gap of understanding with trees, owls, or ducks. Or maybe we haven't found the right way of asking the questions yet. I don't know. Still, it's a pleasure to be among organisms that don't seem to know regret, guilt, or worry. We are the only ones who seem to suffer from these emotions. Although animal and plant consciousness are mysterious, I realized at the Andrews that I believe, as Wendell Berry does, that they do not tax their lives with forethought of grief. I found it deeply comforting to spend time with nonworrying life forms.

The importance of scale

One day I hiked to the top of Carpenter Mountain on the rim of the Andrews watershed. I was standing up there, sweaty and euphoric, with no other humans discernible in my 360-degree view. I could see the whole Andrews Forest, or I could look to the east at the Cascades, and it suddenly seemed so obvious: finding the proper scale is one of the big secrets to finding solace. The right spatial scale and the right temporal scale. Being in the forest, and in wild landscapes in general, always helps me find this perspective, which is one reason being here among old-growth trees brought me such comfort. Their scale is big, mine is small. It's when I imagine that we humans are giant, all important, all destroying, that I feel the most hopeless.

Seen from a little distance, the losses we are causing

feel less tragic. This doesn't mean I can ever forget that we have gotten into deep trouble on our home planet. But in spite of acknowledging our rapacious growth, our using up the land and water, our alteration of the atmosphere, I found that thinking differently about scale freed me. The more I imagined the time scale of life on Earth, of human-induced climate change, of a sixth mass extinction event, the more room I found for the place of peace I'd been cultivating, neither denying nor despairing, of understanding our times and our predicament with some kind of calm.

My growing sense of acceptance of it all was based on thinking more in evolutionary and even geological time. Also, thinking more about the whole planet, not only the places where catastrophes may now be unavoidable. Imagined future losses seem tragic up close, but in the scale of evolutionary and geological space and time, these losses are part of the development of the planet; Earth will survive, and life will evolve. Humans will find a way to come into balance with our planet, or we won't.

I also realized that scale is a key in the struggle over feeling helpless as an individual. Strangely, the trick here is the opposite of realizing that we are small and short-lived in a big world. Time and again, I've felt helpless as an individual. My logic tells me that in the sea of seven billion people, I can do nothing with my actions—lifestyle changes, voting, writing, teaching, activism—to prevent catastrophic global climate change. But reexamined at shrinking scales, that logic holds less power and less truth. The people who work in this forest, who measure the populations of prey, who monitor the rain and snow and the date of first and last frost, who study and protect the owls here, who leave the forest as untrammeled as possible, they certainly were making a difference in this one magical place. I may have little impact in the world, or the country, or even in Oregon. But what about in my community, at my daughter's school, among my friends and family, or in my backyard? There I was on top of that beautiful mountain, looking across a small piece of the Oregon Cascades. I felt in my heart that individual actions could have ripples here.

It's such a cliché: think globally, act locally. But that day at the top of Carpenter Mountain, I began to think about this cliché in a new way. I knew that I wanted to think globally, and even geologically, but acting locally had tripped me up. It felt necessary and important, but I believed that with climate change, acting locally just doesn't cut it. If it were enough, I could feel content to go off-grid, garden, and join the local sustainability coalition. The small-mammal researchers here could feel content that their work would help safeguard future small mammal communities. I could see up on the mountaintop that human-caused climate change altered the whole equation about what constitutes ethical action. Now, I felt that living with integrity requires us to act on a larger stage too, and that acting locally is not enough anymore. Looking for the larger scale could help me find my way forward.

There is no doubt in my mind that we are in for a wild ride. For many months, I had struggled in the shadow of my growing understanding about what we are doing to our atmosphere and what is to come. During my week in the Andrews, I felt the first inkling of that sense of dread falling away. One day I took a long hike in the old growth and lost myself to sweat and tired muscles and a satisfying sleep. Another day I got to the top of the watershed, and clear skies laid out the eastern skyline of central Oregon's peaks before me. I spent another day poking along Lookout Creek, rolling water-smoothed rocks in my hands and getting hypnotized by the river's music.

By the end of the week, I felt recharged, and my heart and my head had cleared. I realized that every visit to a place like this, every hike into wilderness or camping trip to a special place nourishes me for living well in our times. It's so easy to forget about this reliable way to comfort and salvation in the trenches of daily life. A vital piece of responding well to climate change is finding your own reliable way to lift your spirits, and then making sure you do it regularly. I promised myself to get away into open land once a month, and I managed to keep that promise for a full year. It was obvious that I needed to spend more time with "the peace of wild things," because here, I can rest in the grace of the world, and I too am free.

11 Looking for Solutions

I sometimes come across headlines suggesting that a stunning solution to climate change is lurking just around the corner. Here are a few examples: "Will solar save the planet?" *(The Nation);* "The coming green wave: Ocean farming to fight climate change" *(The Atlantic);* "Study: Simple measures could reduce global warming, save lives" *(Washington Post);* "The biggest climate victory you never heard of" *(Al Jazeera English);* "In search of energy miracles" *(New York Times);* and "Can agriculture reverse climate change?" *(Slate).* Inevitably, the story under the headline describes some trend or idea or initiative that is encouraging and that given enough time and interest could add up to real progress. But it is never a miracle solution, and we don't have that much time.

Bill McKibben has said there is no silver bullet that can arrest global warming, only silver buckshot. I began writing this book when I realized that all the silver buckshot we have is not likely to be enough to avoid catastrophic levels of warming, and nothing I've learned has changed my overall assessment about our chances.

Even so, there are solutions that are worth fighting for because the future is unknown. This chapter explores some actions being taken by governments at all levels and by corporations. My focus is on the United States because it's where I live, and because I believe if the United States acts very quickly and powerfully, others may follow—but if the United States fails to act, there is no hope at all.

Federal actions

On the federal level, Environmental Protection Agency (EPA) regulations and a few other national policies have contributed to recent reductions in U.S. greenhouse gas emissions, which are likely to continue in the years ahead. The most significant federal climate actions have been implemented under the Clean Air Act.

The endangerment finding. The Clean Air Act was passed in 1970. Three decades later, a coalition of states, cities, and environmental groups sued the federal government over its failure to regulate greenhouse gas emissions under the Clean Air Act.[1] After a long court battle, the Supreme Court ruled in 2007 that the act required the EPA to determine whether greenhouse gases threaten human health and welfare, and if so, to regulate them. In 2009, the EPA ruled that greenhouse gases do pose a danger; this ruling is known as the endangerment finding.[2]

This court case launched the EPA's painstaking process of regulating greenhouse gas emissions in the United States. In 2010, the largest emitters were required to start reporting their greenhouse gas emissions.[3] Then the EPA drafted rules limiting emissions from new power plants,[4] factories, and vehicles. New fuel efficiency standards for cars and light trucks were issued in 2012 to lower carbon emissions by 40 percent by 2025.[5] The EPA has proposed efficiency standards for heavy-duty trucks, which account for 20 percent of carbon pollution produced by the transportation sector.[6] New gas wells associated with fracking were required to capture methane beginning in 2015,[7] and the EPA has proposed rules to limit methane emissions from landfills and other sources as well.[8]

In August 2015, the EPA released rules for existing power plants that will affect the 491 operational coal-fired power plants in the United States.[9] These rules, also known as the Clean Power Plan, are designed to reduce carbon emissions from power plants 32 percent below peak 2005 levels by 2030.[10] Power plants are responsible for nearly 40 percent of U.S. carbon emissions.[11]

Ever since the endangerment finding was issued in 2009, some states, members of Congress, and fossil fuel interests have fought to undo it. Numerous legal challenges and political attacks have delayed or weakened some rules but failed to derail them. Once the Clean Power Plan was finalized, twenty-seven states joined a suit to block its enactment,[12] and in early 2016, the Supreme Court blocked its enactment until the court case is settled.[13] Even so, if enacted the Clean Power Plan will not dramatically reduce total U.S. greenhouse gas emissions; one analysis found that the EPA rules will result in a 7 percent decline in U.S. emissions by 2030.[14]

Other federal policies. Aside from the endangerment finding and the regulations that have followed, several other national policies have led to small reductions in greenhouse gas emissions. For example, the Department of Energy issues energy efficiency standards for appliances and equipment that have reduced emissions slightly.[15] The federal wind production tax credit has helped spur the expansion of U.S. wind energy; in the years 2008 to 2012, 35 percent of all new generating capacity in the United

States came from wind.[16] This tax incentive's effectiveness has been hampered by Congress's failure to make it permanent: it has expired several times, and its future is perpetually uncertain.

U.S. greenhouse gas emissions in 2013 from energy generation were almost 10 percent below their peak in 2007 and rose only slightly in 2014,[17] the most recent year for which data is available. Projections suggest the country is on track to meet President Obama's goal of cutting U.S. emissions by 17 percent from 2005 levels by 2020.[18] However, this pace of emissions reduction is much too slow to enable the United States to do its part to stabilize global emissions, which would require achieving at least an 80-percent reduction in emissions by 2050.[19] I haven't seen any credible path to this level of emissions reductions without passing comprehensive federal legislation to put a price on carbon emissions.

Carbon-pricing schemes. The two primary approaches used to price carbon are cap-and-trade programs and carbon taxes. Cap-and-trade programs work by setting an overall cap (or ceiling) on greenhouse gas emissions and then issuing permits to industries for each ton of carbon dioxide emitted up to that cap. Companies can then buy and sell permits to emit carbon at auctions, leading to a price on carbon emissions and a financial incentive to pollute less and less. For cap-and-trade systems to be effective, the cap on emissions must be lowered over time.

The devil is, of course, in the details: where the cap is set, whether initial permits are given away or sold, which polluters are included, who gets the money from the auctions and how it is used, and how quickly and steeply the cap is lowered. Cap-and-trade programs require new regulations and an implementation system and thus take time to set up. Problems such as fraud, permit hoarding, low carbon pricing, and failure to lower the cap have plagued the cap-and-trade system in Europe,[20] but the Regional Greenhouse Gas Initiative in the northeastern United States and California's cap-and-trade law (more on these later) have been credited with reducing emissions while spurring economic growth.[21]

The closest the United States has come to passing federal legislation to regulate greenhouse gas emissions was in 2010, when a bipartisan cap-and-trade bill nearly reached a vote but then died in the Senate. Since then, with the rise of the Tea Party, opposition among Republicans in Congress to such a law has become virtually unanimous, and Democrats have soured on the approach as well.

A carbon tax is a direct fee on each ton of greenhouse gases emitted to discourage activities that emit carbon and encourage alternatives and conservation. If the tax is set high enough, renewable energy and efficiency become cheaper than fossil fuel energy, leading to conservation, innovation, and a conversion to clean energy. Because the purpose of a carbon tax is to reduce carbon emissions rather than to fund government, the money collected can be used to reduce other taxes or returned to households to offset higher energy prices. A carbon tax that returns all funds to taxpayers is called a carbon fee and dividend or a revenue-neutral carbon tax.

As with a cap-and-trade program, the details of a carbon tax determine its impacts and effectiveness. The most important details are the size of the initial tax, whether and how much it increases over time, and what is done with the collected money. To be effective, a carbon tax needs to be high enough to change the behavior of individuals, businesses, and industries. What is done with the collected money determines who is harmed and who benefits, but any sizable carbon tax would be effective in reducing greenhouse gas emissions. A 2014 analysis examining the impacts of a revenue-neutral carbon tax in the United States starting at $10 per ton of CO_2 and rising $10 per ton annually projected that over a twenty-year period, CO_2 emissions would be reduced 50 percent below 1990 levels.[22]

Momentum seems to be building for taxing carbon emissions. Many leading climate scientists, conservative and liberal politicians, economic modelers, public opinion experts, and world leaders have come out in support of taxing carbon. In Canada, British Columbia enacted a revenue-neutral carbon tax in 2008 that is extremely popular and has reduced fossil fuel use while boosting the economy,[23] and Alberta recently announced plans to implement a revenue-neutral carbon tax as well.[24] According to a 2015 report by the World Bank, forty national and twenty subnational jurisdictions have implemented or are planning carbon-pricing schemes, including Iceland, Sweden, Norway, France, Portugal, and Mexico.[25] At the Paris Climate Conference in December 2015, six heads of state along with prominent business and banking leaders launched the Carbon Pricing Leadership Coalition to promote the use of carbon pricing around the world.[26] In the United States, several 2015 polls suggest fully refunded carbon taxes may find broad public support.[27] The Carbon Tax Center provides technical information and analysis in support of carbon taxes, and Citizens' Climate Lobby (CCL) is working to pass national legislation to implement a revenue-neutral carbon tax.

Local, state, and regional actions

Many communities, states, and regions in the United States are taking action to bring down greenhouse gas emissions. Renewable energy standards, cap-and-trade programs, clean energy tax credits, energy efficiency projects, and other strategies are in place in much of the country. For example, thirty states and the District of Columbia have established renewable energy standards of some sort. These range in ambition: North Carolina's renewable energy standard (RES) calls for 12.5 percent renewable energy generation by 2021, while California's RES calls for 33 percent by 2020.[28] In June 2015, two states surpassed any previous RES ambitions: Hawaii passed a law aiming for 100 percent renewable electricity generation by 2045, and Vermont committed to 55 percent renewables by 2017 and 75 percent by 2032.[29] In early 2016, my home state of Oregon passed a law mandating that our two largest utilities eliminate coal power by 2030 and reach 50 percent renewable energy by 2040.[30]

In 2005, ten northeastern states established a cap-and-trade program called the Regional Greenhouse Gas Initiative (RGGI), which includes only emissions from power plants. Emissions auctions began in 2009, and currently, RGGI states include Connecticut, Delaware, Maine, Maryland, Massachusetts, New Hampshire, New York, Rhode Island, and Vermont (New Jersey dropped out in 2011). Although the program initially struggled with low prices and inadequate demand for CO_2 allowances due to the recession and cheap natural gas, it has since had some success.[31] The cap was lowered significantly at the end of 2013, causing the price of allowances to rise quickly.[32] A 2015 analysis concluded that the RGGI has been an important factor in reducing greenhouse gas emissions, and without it, current emissions would be expected to be 24 percent higher than they are today.[33]

State investments and incentives have had an impact in some states as well. New Jersey's Renewable Energy Tax Credit program was highly effective in spurring solar installations, which went from close to zero in 2005 to near 16,000 in 2011. By the end of 2011, New Jersey had the second largest solar market in the country, with an estimated $1.2 billion in investments, 800 megawatts of solar capacity, and 10,000 jobs supported by the solar industry.[34] After the incentives were removed, New Jersey fell from the second to the seventh largest solar market in the United States.[35] Texas, with its vast expanses of windswept prairie, leads the United States in wind power production and generated more than 10 percent of its electricity from wind in 2014.[36] The wind boom was facilitated by the state of Texas investing $7 billion in 2005 to build 3,600 miles of transmission lines to get the electricity from wind farms in west Texas to major population centers.[37]

A number of U.S. cities are taking action as well. New York City has set a goal of reducing emissions 40 percent by 2030.[38] San Francisco developed a climate action strategy that has reduced carbon emissions 14.5 percent below 1990 levels and aims to reduce emissions 40 percent below 1990 levels by 2025.[39] Palo Alto, California (pop. 66,000), committed in 2013 to 100 percent carbon neutral electricity, the first U.S. city to do so.[40] In 2015, a coalition of municipal, state, and international governments set a goal of reducing greenhouse gas emissions 85 to 90 percent by 2050, or to less than 2 metric tons per person. This group includes Vermont, California, Oregon, and Washington, as well as Brazil, Canada, and Germany. This goal is a science-driven target aimed at limiting warming to 2 degrees C.[41]

And then there's California, with the eighth largest economy in the world.[42] In 2006, California passed AB 32, the Global Warming Solutions Act. Signed by then-governor Arnold Schwarzenegger, the law committed the state to reducing greenhouse gas emissions to 1990 levels by 2020, with a target of an 80-percent reduction by 2050.[43] In 2015, Governor Brown issued an executive order aiming at an interim 40-percent reduction in emissions by 2030.[44] These targets make California, as far as I know, the only major world economy that has laws in place to reduce emissions quickly and steeply enough to do its part to contain global warming to 2 degrees C.

California's law authorized a cap-and-trade program, which took a number of years to construct and implement. The first emissions auction was held in November 2012 and included power plants, refineries, and cement plants. Auctions are held quarterly, and tailpipe emissions were added in 2015.[45] So far, AB 32 has survived multiple legal challenges and is supported by a majority of California voters.[46]

California is on track to meet its RES goal mandating that 33 percent of electricity be generated from renewables by 2020.[47] In 2014, California doubled the amount of solar energy flowing into its grid in a single year[48] and became the first state to generate more than 5 percent of its electricity from the sun.[49] Wind power capacity has doubled since 2009, and in 2015 wind provided about 7 percent of California's electricity.[50] California's Low Carbon Fuel Standard (LCFS) has helped the state begin transforming its vehicle fleet to dramatically lower transportation emissions.[51]

Corporate actions

In the absence of strong federal action, some corporate decision makers are growing alarmed about the potential impacts of climate change on their businesses and taking steps on their own to reduce emissions. IBM, FedEx, and WalMart each have made commitments to cut emissions through energy efficiency and energy upgrades, waste reduction, changing their supply chains, and other measures.[52] Coca-Cola has spoken out about climate disruptions such as droughts and floods affecting access to the water and sugar the company needs for its products.[53]

Many insurance industry giants are warning about increasing climate risks, breaking with industry groups long dominated by energy companies promoting climate change denial.[54] In December 2014, 221 major companies—including Nike, IKEA, Nestle, Starbucks, and Levi Strauss—signed a letter in support of the EPA's Clean Power Plan.[55] It has also been reported that some of the largest U.S. corporations will accept and are planning for a carbon tax.[56] According to CDP, a nonprofit working with businesses to tackle climate change, 435 large companies had established an internal price on carbon emissions by 2015, and 583 anticipated doing so in the next two years.[57] Ahead of the Paris Climate Conference in December 2015, a group of European oil companies (including Shell, Total, and BP) urged the adoption of effective carbon pricing,[58] and six major U.S. banks called on global leaders to adopt a strong climate agreement.[59]

The Paris climate accord finally put the world's countries on record as agreeing to work together to fight climate change.[60] However, the voluntary pledges submitted in Paris, even if they are fully reached, are only about half of what is needed to have a good chance of holding warming to the 2-degrees-C goal.[61] We know this because data exist on exactly what it will take to contain emissions to that quota. A report from PricewaterhouseCoopers called *Low Carbon Economy Index 2015* found that in 2014, the world's major economies were reducing their carbon intensity (a measure of greenhouse gases emitted per unit of economic growth) at an annual rate of 1.3 percent. According to the report, in order to meet the target of holding warming to 2 degrees C, we need to have an annual reduction of at least 6.3 percent every year from now until 2100.[62] As noted in "Starting Premises," recent evidence suggests 2 degrees C of warming may trigger severe and irreversible climate effects, leading to calls for a 1.5-degrees-C limit on warming.[63]

The math is strongly against us. So is time. However, getting there is not yet completely impossible.

12 Spotting Hopeful Trends

What do all the efforts described in the previous chapter mean? I don't know. In 2011, I would have said they probably mean nothing, but recent signs of progress make me more hopeful. The slowing of tropical deforestation, the decline of coal, the increasing accessibility of renewable energy, the rise of climate activism, the passage of the Paris climate accord all suggest that change is possible. This chapter chronicles some of the most hopeful developments I've discovered.

The slowing of tropical deforestation

The Union of Concerned Scientists estimates that about 10 percent of our greenhouse gas emissions today come from forest clearing and forest degradation.[1] But this estimate represents real progress; in the 1990s, tropical deforestation was destroying sixteen million hectares of forest a year and was estimated to contribute 17 percent of total global warming emissions.[2]

Brazil and Indonesia, the two countries with the

largest expanses of tropical forest cover, were losing forests rapidly throughout the 1990s. Many countries, but especially Brazil, saw steep declines in deforestation rates in the first decade of the twenty-first century, leading to an estimated 19-percent drop in the tropical deforestation rate worldwide.[3] India, one of the most densely populated countries in the world, began to reverse its deforestation trend in the 1990s, and today its forests are sequestering carbon from net forest regrowth.[4] A 2010 analysis of data on worldwide carbon emissions found that land use CO_2 emissions had declined from an estimated 5.32 billion tons per year during the period from 1960 through 1999 to an estimated 3.23 billion tons in 2009.[5]

This rare climate success story is actually many stories. Slowing deforestation in tropical countries in the 2000s was achieved in different ways in different places, but involved a lot of hard work, determination, and money. One successful approach, in which developed countries pay tropical countries to reduce emissions through protecting forests, is a United Nations program called Reducing Emissions from Deforestation and Forest Degradation in Developing Countries (REDD+).[6] The REDD+ program has been credited with helping to slow deforestation in Guyana, Brazil, Costa Rica, Kenya, and Madagascar in recent years.[7]

Much of the tropical deforestation of the 1990s was driven by the expansion of industrial-scale soy, cattle, and palm oil cultivation in tropical countries. In 2006 a series of consumer-focused campaigns and boycotts began to publicize the destruction of tropical forests by the beef, soy, and palm oil industries.[8] These campaigns focused initially on Brazil and resulted in a number of the largest corporate beef and soy purchasers pledging to exclude Amazon deforesters from their supply chains.[9] Next came

Indonesia and Malaysia, where palm oil cultivation has destroyed at least 30,000 square miles of tropical forests.[10] The pressure has gotten results: in 2012, twenty of the world's largest companies made commitments to eliminate deforestation-causing products from their supply chains.[11] In 2013, Asia Pulp and Paper, responsible for much of the deforestation in Sumatra, made a pledge of zero new deforestation of natural rainforests.[12] In 2014, four of America's largest palm-oil consumers–Colgate-Palmolive, General Mills, Kellogg, and Proctor and Gamble—pledged to use only deforestation-free palm oil,[13] and Cargill, one of the world's largest agricultural companies, committed to a zero-deforestation pledge for all the commodities it produces.[14] It's too soon to know if these pledges will be honored; Indonesia continues to be a world hotspot for deforestation,[15] but modern satellite systems will allow compliance to be verified.[16]

Other actions by both individuals and governments have contributed to slowing tropical deforestation. Expansion of protected areas, indigenous reserves, and sustainable-use forests have been important in some countries, including Brazil and Madagascar. National legislation, better enforcement of existing laws, and community forestry projects have yielded benefits in India and elsewhere.[17] The Congo Basin in central Africa contains the second largest area of rainforest in the world after Amazonia. This vital carbon sink has maintained extremely low deforestation rates through a combination of government policies, inaccessibility to industrial agriculture, and urbanization trends.[18]

In some tropical countries, cleared or damaged lands are regrowing forests due to changing societal or economic conditions.[19] There seems to be no consensus yet about how much land this involves or how important it is for the climate, but tropical forest regrowth has been documented in Costa Rica,[20] Vietnam,[21] the Amazon basin, and El Salvador.[22] Meanwhile, the future direction of the Brazilian Amazon's forests is uncertain; deforestation rates there fell in the 2000s but are reportedly again on the rise.[23]

While it's not clear whether recent successes in slowing deforestation along with forest regrowth will result in net carbon absorption by tropical forests anytime soon and whether rising tree mortality rates documented around the world will cancel out these potential gains, the slowing of tropical deforestation shows what can be done with focused effort.

The decline of coal and the rise of renewable energy

Climate experts disagree on a list of energy topics that includes nuclear power, hydraulic fracking, expansion of hydroelectric power, and whether renewables can fully power civilization, but they do agree that most of the world's remaining coal cannot be burned if catastrophic levels of warming are to be avoided.[24] In 2005, 50 percent of U.S. electricity generation came from coal,[25] but burning coal in the United States has been in decline since then.[26]

Many energy experts consider the decline in coal burning to be this country's most significant action to avert climate change so far.[27] One factor in coal's decline has been the rise of domestic natural gas production from fracking.[28] Many other factors—the economic crisis, increased conservation and energy efficiency, growth in renewable energy production, and tighter EPA regulations for coal plants—have contributed to coal's decline as well.[29] Finally, the Sierra Club's Beyond Coal campaign has been highly effective in preventing new coal plants from being built in the United States and shutting down the oldest and most polluting plants as well (more on this later).[30]

Coal's decline is projected to continue.[31] In 2015, a record 7 percent of U.S. coal-burning capacity went offline as tightened EPA mercury emissions standards took effect and other factors also encouraged the retirement of older coal plants.[32] Grassroots efforts have defeated nearly every proposed new coal plant in recent years,[33] and if the Clean Power Plan is not derailed, it will likely prevent new coal plants and speed the retirement of existing ones.[34]

As coal use has fallen, renewable energy generation has been expanding. In 2014, the most recent year for which data is available, renewable energy accounted for 10 percent of U.S. energy consumption and 13 percent of electricity generated.[35] That same year, solar energy accounted for 32 percent of all new generating capacity in the United States.[36] Leading states included Texas, generating about 10 percent of its electricity from wind, and California, producing more geothermal and solar energy than any other state.[37] With this expansion, costs have fallen dramatically, and solar technology has rapidly been gaining efficiency. According to a 2015 report, the cost of wind power fell 61 percent from 2009 to 2015, and the cost of solar fell 82 percent. At 2015 costs, under some scenarios, electricity generation from solar and wind is now cost competitive with fossil fuel electricity.[38]

In many other parts of the world, shifts toward renewable energy are beginning as well. The U.S. Energy Information Administration estimated in 2014 that 11 percent of world energy production was coming from renewable sources.[39] In 2014, China increased its spending on renewables by 33 percent to $83 billion; globally, $270 billion was reportedly invested in renewables.[40] In 2015, a group of some the world's richest tech leaders, including

Bill Gates (Microsoft), Mark Zuckerberg (Facebook), and Jeff Bezos (Amazon) launched a new investment project named the Breakthrough Energy Coalition to fund the development of renewable energy technologies.[41]

The question of whether it is still possible to shift to renewable energy and cut back consumption quickly enough to hold climate change to manageable levels is hotly debated. It is hard to imagine a shift from 10 percent to close to 100 percent renewable power generation in a few decades, along with the abandonment of all our fossil-fuel power and infrastructure. Overcoming the intermittent nature of solar and wind power generation will also present huge challenges. In 2015, a team of Stanford researchers produced a detailed road map outlining how the United States could convert to 100 percent renewable energy by 2050 with current technologies and strong economic benefits.[42] This remaking of our energy, transportation, and infrastructure in just a few decades has been called impossible by other energy experts, who have criticized the Stanford work as flawed and unrealistic.[43]

If it is possible it will require a massive and rapid rise in popular, political, and social demand and widespread commitment to making this change, and soon.

The rise of climate activism

Climate activism is on the rise both in the United States and globally. I squirm a little as I write this, because it's difficult to prove, and yet I'm sure it's true. Many U.S. environmental groups tried to take on climate change in the 2000s and worked for passage of a federal cap-and-trade bill. When that legislation stalled in 2010, there was no popular outcry because its passage had never had strong grassroots support. Since then, many of these environmental groups have made combatting climate change a central goal, and grassroots campaigns focused on climate are growing.

According to a 2014 analysis, the ten largest groups with climate as a major priority had fifteen million members and combined annual budgets of more than $525 million.[44] Two new groups devoted exclusively to restoring a stable climate have been founded in the United States: Citizens' Climate Lobby in 2007 and 350.org in 2010. Both are growing and mostly powered by volunteers. Several targeted yet highly successful campaigns have been launched to bring down greenhouse gas emissions and stabilize the climate. What follows are short descriptions of what I see as the five most important and promising climate crusades in the United States today.

The Beyond Coal campaign. Spearheaded by the Sierra Club, Beyond Coal has worked to replace coal with clean energy one community at a time. Since 2010, with funding from Michael Bloomberg, Beyond Coal has partnered with local activists to mobilize communities and mount legal challenges to coal plants across the United States.[45] By focusing on the devastating human heath effects, high carbon emissions, and economic risks associated with coal burning, Beyond Coal activists have been extremely effective: according to their website, 230 coal-burning power plants have been retired in the United States so far, and 189 proposed plants have been defeated before being built.[46] According to the Sierra Club, this reduction in coal-fired power has helped the U.S. electricity sector reach its lowest carbon emissions since 1995.[47]

Halting coal exports. A ton of coal burned anywhere on Earth emits the same amount of CO_2, so switching off coal in the United States is pointless for the climate if that coal is mined, exported, and burned elsewhere. With domestic coal use declining, coal companies have tried to vastly expand exports of U.S. coal overseas, especially to Asia.[48] However, to expand coal exports, these companies require vastly expanded export infrastructure for coal trains and barges, along with coal export terminals. They have been trying to construct infrastructure for exports of Wyoming coal through the Pacific Northwest, and for exports of Appalachian coal through Louisiana and Texas. Dozens of grassroots campaigns have burst forth to fight these projects and have been extremely effective.[49]

In the Pacific Northwest, activists fighting coal export terminals and other fossil fuel export infrastructure have helped derail plans for three of six proposed coal export terminals so far, and a fourth proposed terminal is facing fierce protests from climate activists and Columbia River tribes.[50] In December 2014, a proposed coal export terminal in Louisiana lost its permit in state court after facing strong community opposition.[51] Out of fifteen proposals to build new export terminals on U.S. coasts, eleven were defeated or canceled in 2013 and 2014 alone.[52] These victories are not the end of the battle over U.S. coal exports, but they do illustrate how effective climate activists have been in delaying or thwarting U.S. coal export expansion plans.

The No KXL campaign. The Keystone XL pipeline system, first proposed in 2005, was supposed to carry at least 800,000 barrels of tar sands oil per day from Alberta, Canada, to Texas refineries[53] but required a federal permit to cross the U.S.-Canada border. Climate activists led by 350.org united to fight this pipeline, citing the terrible destruction and pollution caused by tar sands mining, the risk of oil spills and water pollution along the pipeline route, and high greenhouse gas emissions from the extracting, refining, and burning of this particularly high-carbon form of oil. Protesters demonstrated at the White House, along the proposed route, and around the country, and it's pretty clear that this campaign helped galvanize the environmental movement to fight climate change. Native American tribal members, farmers, and ranchers teamed up with climate activists to fight the pipeline, and this coalition seemed to strengthen over time. James Hansen, Bill McKibben, and hundreds of others were arrested at the White House protesting against the pipeline.

The permit application was first filed in 2008, but the No KXL campaign helped delay the decision multiple times while it was under federal review.[54] President Obama ultimately rejected the request for the permit in November 2015, citing climate change as his reason.[55] This was a huge win for climate activists, whose opposition to Keystone early on was not shared by Congress, the White House, or the majority of Americans, and who were never expected to win this fight.

The fossil fuels divestment movement. In November 2012, Unity College, a tiny college in rural Maine, became the first college in the United States to divest its endowment of all fossil fuel holdings. Since then, the divestment movement has spread to many campuses, cities, religious institutions, and foundations. The goal is to get institutions to sell all investments in fossil fuel companies, particularly the two hundred companies that hold the vast majority of coal, gas, and oil reserves. This campaign is coordinated by 350.org.

The movement isn't attempting to cause economic harm to fossil fuel companies. Instead, it is attempting to change the moral and political thinking of society, to make the statement that "it is wrong to ruin the climate, and because the fossil fuel industry is committed to ruining the climate, it must lose its license to operate."[56] The divestment movement has had many successes over the last several years: more than fifty colleges and universities, including the University of Hawaii, Stanford University, and the University of California system, as well as cities, foundations, religious groups, and other institutions in the United States and around the world have committed to divest.[57] Several large institutional investors have pledged to divest, including the Church of England ($14 billion investment fund), Norway's government pension fund ($890 billion),[58] and California, affecting $476 billion in retirement funds.[59] As of December 2015, more than five hundred institutions, responsible for $3.4 trillion in investments, had made some kind of divestment commitment.[60]

The carbon tax campaign. Many prominent economists across the political spectrum advocate taxing carbon emissions as the most efficient, most effective, and cheapest way to reduce greenhouse gas emissions.[61] Citizens' Climate Lobby (CCL) is working to educate the public about this solution, lobby elected officials, and pass national legislation implementing a revenue-neutral carbon tax. The CCL proposal is for a greenhouse gas emission tax that starts low and rises steadily, with all money collected returned equally to households to offset the higher cost of energy during the transition off fossil fuels.

Started by a few people in San Diego in 2007, CCL has grown to tens of thousands of volunteers in three hundred and ten chapters across the United States, Canada, and around the world. Chapters are in every state and most U.S. congressional districts to educate and lobby. Climate activists have also launched campaigns to pass state-level carbon taxes in Massachusetts, Washington, Oregon, Rhode Island, and Vermont.[62]

Citizens' Climate Lobby carbon fee and dividend proposal:

- A fee would be placed on carbon-based fuels at the source (well, mine, port of entry). The fee would start at $15 per ton of CO_2 emitted and increase steadily each year by $10 per ton. Within a decade, clean energy would be cheaper than fossil fuel energy.
- All revenues would be distributed equally to American households as a monthly dividend check to offset higher energy costs.
- A majority of households would break even or receive more in their dividend than they would pay in higher prices due to the carbon fee, protecting the poor and middle class from economic harm.
- According to a study by Regional Economic Models Inc., this proposal would reduce greenhouse gas emissions in the United States 50 percent below 1990 levels in twenty years while creating 2.8 million new jobs.
- Because the plan returns the collected fees, it would not raise taxes or grow the government, making it a nonpartisan proposal.
- Border tariffs would be applied to imports coming from countries that do not have an equivalent carbon price, protecting American businesses and providing other countries with an incentive to price carbon.
- Individuals would have a financial incentive to reduce carbon emissions, and businesses would have certainty about the future price on carbon emissions.[63]

In 2015 and early 2016 there were many signs of progress and many firsts. Leadership changes in China, Canada, and Australia moved those countries toward serious efforts to address climate change. Canada announced plans for national carbon pricing.[64] President Obama's decision on the Keystone XL pipeline was the first rejection of a major fossil fuel infrastructure project on the basis of its climate impacts. Eleven Republican members of the House of Representatives introduced a resolution recognizing humans' role in climate change and committing to work for solutions.[65] Several senators introduced a bill to end the leasing of federal lands for fossil fuel extraction—a proposal inconceivable even a few years ago.[66] Although the Paris climate agreement of December 2015 was criticized as too modest to prevent catastrophic warming,[67] the accord represents the first time nearly every country on Earth agreed to fight climate change together. In early 2016, two representatives (a Republican and a Democrat) launched the bipartisan Climate Solutions Caucus in the House of Representatives.[68] A group of attorneys general from 18 states and territories formed a coalition to investigate ExxonMobil and other energy firms for misleading their investors and the public about the reality and risks of climate change for decades.[69]

Finally, an unprecedented legal campaign known as Atmospheric Trust Litigation[70] scored two huge victories in Federal and State courtrooms. In 2011, the non-profit group Our Children's Trust launched climate litigation on behalf of youth and future generations seeking a legal right to a healthy atmosphere and a stable climate.[71] This campaign has launched legal proceedings in every U.S. state, sued the federal government, and provided support for attorneys in a dozen other countries. Many of the early cases were dismissed, but in April, 2016, a federal judge in Eugene, Oregon, ruled that a case brought by 21 youth suing the Obama administration could go forward.[72] And in a surprise ruling, a judge in King County, Washington, ruled that the State of Washington must produce emissions reduction rules by the end of 2016 and make science-based greenhouse gas emissions reduction recommendations to the legislature in for the 2017 session.[73] These rulings open the possibility that some courts may be ready to recognize that the government has a duty to protect the climate to preserve a habitable planet for young people and future generations, and that the government is failing.

To me, these were signs we may be on the verge of truly responding to—rather than denying or ignoring—climate change.

13 Preparing

When we first moved to Oregon, I was a little put off by all the women I met who knew how to can fruit, make pickles, spin wool, and prepare every cut of meat from a lamb—not to mention the weavers, master gardeners, sewers, cheese makers, bread bakers, quilters, and carpenters. I was intimidated, I guess, but it also seemed a little backward to me. I had just finished writing my dissertation and then struggled through a difficult pregnancy while doing ecological fieldwork; when we moved here with our newborn daughter, I didn't think I wanted to learn homesteading skills. But my attitude has changed.

In fact, I have become convinced that rebuilding some self-sufficiency is part of the way forward in the face of climate destabilization. I've also discovered that doing more for myself is surprisingly satisfying. Clearly, I'm not alone. Reweaving nearby self-sufficiency systems is at the heart of many growing trends in the United States and around the world—relocalization, gardening, backyard chicken and beekeeping, and efforts to relearn food processing and preserving skills. I've even seen workshops offered to teach people how to butcher their own animals. Home gardening, community gardens, and urban farming projects have recently surged across the United States.[1] In many regions, buy-local campaigns are blossoming, with a strong emphasis on rebuilding local economies.[2] According to the USDA, an increasing number of farms, nearly 8 percent in 2012, are selling food directly to consumers.[3]

Building resilience

Like *sustainability, resilience* is one of those words that refers to something important but is difficult to define and becoming overused. I use the term to mean the strengthening of our ability to withstand and bounce back from unexpected setbacks and harms. Building resilience includes safeguarding ourselves by making basic preparations, relearning how to do more of what it takes to survive and thrive, and connecting personally with people who have skills we lack.

I, like many others, fear the possible breakdowns, the end of the era of plenty, and all the unknowns that will come with accelerating climate change. I have the sense that more skills will give me more options for responding, and I think this is behind many recent D.I.Y. and relocalization efforts. My friend Theresa told me she has developed an unexplainable urge to learn how to make pickles. I think it's because it makes us feel safer to know how to *do* things.

People in many communities in the United States and around the world are starting projects promoting resilience and self-sufficiency. One of the early manifestations of this was the founding of the Transition Movement in 2006 by a British permaculture instructor named Rob Hopkins. His premise is that peak oil and climate change will soon cause the collapse of modern industrial civilization and that the best response is to start preparing by building local resilience, self-sufficiency, and low-carbon lifestyles now. Many others besides Hopkins have predicted collapse, but they generally predict a dark and dystopian future that will be about doing without modern comforts and learning to survive in a harsh world.

The Transition Network has developed a different storyline. It does begin with the premise that our current way of life, dependent as it is on fossil fuels and a stable global climate, is not going to last much longer. But Hopkins's idea is that "if an entire community faces this stark challenge together, it might be able to design an 'elegant descent' from (peak oil)."[4] This path into a lower-energy lifestyle of walkable villages, local food systems, and greater connections with the natural world, Hopkins believes, will lead to richer and more enjoyable lives than what we have now.

The Transition Movement approach is to just go out and start something that will increase the self-sufficiency and resilience of your community, build relationships, and be fun. Anything goes: sewing lessons, tree planting, community gardens, local energy production, local currencies, energy conservation projects, local food co-ops, oil memorials. Transition philosophy is that as people see it's possible to prepare a little for the unknown, and as they learn they are not alone, they begin to be able to take action, have fun, and imagine a different future.[5] According to the Transition Network website (transitionnetwork.org), there were 1107 Transition communities in 43 countries in 2013,[6] and many more communities were in the "mulling" stage of developing Transition projects.

The possibility of a different future, right here

I don't believe the Transition Movement is going to take over the world, or prevent or protect us from climate change, but what I like about it is the imagining of a future different from and much more joyful than the more typical dystopian narrative.[7] The possibility of a different future is what I've been searching for all along, and becoming involved in local resiliency projects that fit the Transition mind-set has helped. These projects—part preparedness, part skill building, and part networking—are probably going on in your community too.

Here in the Willamette Valley, farmers' markets are thriving and now run year-round, CSAs are on the rise, local foods are labeled and promoted in a number of grocery stores and restaurants. A group of growers and businesses have established the Southern Willamette Valley Bean and Grain Project to build organic staple food production and processing infrastructure. They have reestablished local wheat and grain production, set up storage infrastructure, and built two commercial grain mills. All-local products including oatmeal, beans, flour, and bread are now being produced and sold here. Local pastured beef, lamb, pork, chicken, and eggs are widely available year-round. The farmers' markets here accept SNAP cards (formerly food stamps), and a local food group offers incentives doubling SNAP benefits for customers using them at farmers' markets.

The Corvallis Sustainability Coalition, made up of community volunteers and more than two hundred partner groups, has worked on a variety of initiatives to make Corvallis more sustainable, less wasteful, and more resilient. Members work on rainwater harvesting, alternative transportation, walkable neighborhoods, renewable energy, waste reduction, and a variety of other projects. Coalition programs have helped establish community gardens, promoted local businesses and farms, and led to instituting fareless buses throughout Corvallis. A project called Take Charge Corvallis is aimed at dramatically reducing energy use from utilities in our community through education, outreach, and incentives. In the event of disruptions to the supply of food, fossil fuels, or other needed goods, these projects will provide some buffering.

Looking around our home at the backyard fruit trees, the bag of local wheat, and the freezer full of food we've gathered, harvested, and bought from nearby, I realize how many of the skills I first wanted to avoid I've learned since our move to Corvallis in 1999. One year my friend Karen taught me how to make jam from the plums in our backyard. Another year, Theresa and I made mozzarella cheese. I learned to bake bread, started knitting again, and learned how to prune our fruit trees. One year I bought tuna from a fisherman on the coast, borrowed a smoker, and learned how to smoke fish. I shop year-round at the farmers' markets, and we continue to add to our yard: an herb garden, raspberries, an apple tree, blueberry bushes. Last year, Mark and I finally succumbed to a western Oregon fall ritual and gathered wild chantrelle mushrooms for the first time.

Why prepare?

Asked to predict what the future holds, Wendell Berry said, "How can so many people of certified intelligence have written so many pages on a subject about which nobody knows anything?"[8] He has a point; we can't really *know* anything about the future for sure. Even so, preparing for disruptions to the global climate and the global economy is a piece of responding to climate change. Learning skills I used to believe I didn't have time for makes me feel safer. And buying from local farms, food processors, businesses, and craftspeople makes me feel safer too.

I've come to imagine these efforts to build skills and resources in my community as putting together a rudimentary safety net, like the frame of an unfinished house, one that with more people participating could grow into a very basic shelter. If so, the thing to do is to get involved: learn something you can do yourself and buy from nearby producers. These actions won't prevent the melting of the ice sheets or stop the increasingly severe floods and droughts around the world or prevent famines or the extinction of polar bears. They won't be enough to bring down greenhouse gas emissions to levels that will prevent catastrophic climate change.

But my efforts to learn skills and build local sources for some necessities have helped me in both tangible and abstract ways; I feel a little bit safer but also have a richer and more connected life. I can imagine the possibility that the future may not just be difficult or tragic but may be good too. This matters because as much as we can't know the future, we imagine it all the time, and how we imagine it changes how we feel and how we act.

14 Finding My Way

When I began writing this book, everything I learned about climate change felt like a new piece in a heavy, invisible chain-link necklace I wore. I continued, hoping to arrive at a different way of experiencing our current moment on Earth. Although I can't say exactly when or how it happened, my perceptions did change quite dramatically. Partly, this was a personal transformation, but the world has changed too. Today, the way I experience climate change is as a shimmery hologram that I can look at from various angles. From each perspective, I see something different.

The tipping point angle—hopeful shifts

Remember that in 2006, climate scientist James Hansen began saying something pretty alarming. Based on his analyses of ancient climates, emissions trends, climate feedbacks, and climate models, Hansen predicted that we were nearing a climate tipping point. He estimated that we had about ten years to reverse rising greenhouse gas emissions trends; he said the data suggested that if we didn't, rapid, catastrophic, irreversible climate change would ensue.[1] At that time, Hansen was NASA's top climate scientist and the leading scientific authority in the United States on climate change; reading his warning had a profound impact on me. I wrote an essay about Hansen's prediction, and I carried it around my neck as one of the first links in my necklace—and perhaps the heaviest.

In the years since Hansen issued this warning, I've noticed three important shifts. The first is that he has stopped discussing a climate tipping point and become an outspoken and tireless climate activist. He has retired from his post at NASA to speak, write, lobby, do research, and protest full time for action to combat climate change.[2] The second shift is that the signs Hansen warned about in his tipping-point presentations—disruption of global weather patterns, increases in flooding, melting of polar ice caps, droughts, heat waves, wildfires, and species extinctions—have all increased. And the third change is that I've begun to see signs of the possibility of a different kind of tipping point—not physical but cultural.

I don't know for sure why Hansen no longer discusses a climate tipping point, but my guess is that he decided it was too discouraging and wasn't spurring people to take action. Hansen has been right about every major aspect of climate change since he began working on it in the 1970s, and I believe he's right about an approaching climate tipping point too. Hansen has endured criticism from his peers for speaking out, he's been arrested, and he's been threatened. I'm pretty sure he still fully believes that we are passing a critical juncture with greenhouse gas emissions and the global climate system; he's just decided not to talk about it in those terms.

When Hansen first discussed a climate tipping point, a few people (including me) became deeply alarmed, but no groundswell of public understanding or action followed. There was simply too much resistance, denial, ignorance, confusion, and apathy. Fossil fuel and corporate interests did their best to spread misinformation and sow confusion, and they were highly effective.[3] But as the climate destabilizes, I've begun to see the possibility of the public reaching a tipping point. This shift will occur if and when a critical number of us understand that rapid, human-induced climate change poses an imminent threat to everything we as a species need, depend on, and love. If this time is near, I believe its approach began with the 2011 string of unparalleled major weather disasters in the United States. Since then, the extreme weather in the United States and around the world that has continued to wreak destruction and set records seems to have gotten people's attention.

I've detected, among my friends and acquaintances, a shift in understanding; I've seen it in the most recent polls and in the media too. It seems to be dawning on people that global warming is not a political issue but actually a fact. On September 21, 2014, the very first worldwide demonstration for action on climate change was held, with an estimated 400,000 people marching in New York City and 2,800 concurrent solidarity events in 166 countries.[4] In 2015, we saw the many signs of change discussed in Chapter 12. In early 2016, even as the GOP presidential candidates continued to evade discussing facts, two Florida members of the House—one Republican and one Democrat—created a bipartisan climate solutions caucus to look at options to address climate change.[5]

I don't know if we are there yet with climate change, but cultural tipping points of this kind can come suddenly. The fall of the Berlin Wall, the Arab Spring, the end of apartheid, the rapid acceptance of gay marriage, the legalization of marijuana—all of these huge tipping points came as surprises to me. I am confident we are much closer than we were in 2006 when Dr. Hansen issued his initial warning, and when the climate still seemed relatively predictable. The next few years will tell, but it feels possible we're about there.

The resistance to change angle—failure is likely

Another angle on climate change began to crystalize for me while I was talking to my mom. She's a retired psychologist and is endlessly interested in people in a way that I'm just not. One day, I was ranting to her about the seeming insanity of climate inaction as the evidence of dangers and impending catastrophes piled up. She was just listening, but then she sighed and said, "Well, people really hate change."

I had thought a lot about the effectiveness of the energy industry in blocking, stalling, and derailing action. Even so, I had trouble understanding why those of us with nothing to gain from Chevron and Exxon-Mobil's record profits were doing so little. Of course, if you let yourself become engaged, one thing leads to another. You have to change. My mom's words really landed and stuck with me. I know a few people who thrive on change, but as I looked around I realized that she was right: most of us live our lives trying to avoid it.

Bill McKibben wrote this about the difficulty of choosing change: "Most of us are fundamentally ambivalent about going green: We like cheap flights to warm places, and we're certainly not going to give them up if everyone else is still taking them. Since all of us are in some way the beneficiaries of cheap fossil fuel, tackling climate change has been like trying to build a movement against yourself."[6]

Entire books have been written about why we have failed so completely to face and tackle the threat of global warming. In fact, it's a bit of a mini-fad to discuss this topic and to offer ways of reframing, reimagining, or reworking the problem to overcome our resistance to dealing with climate change.[7] While I think these analyses have value, for me the take-home message is simple, and Bill McKibben and my mom summed it up. Given that we are all complicit in a carbon-emitting economy, and that we (as a gross generalization) hate change, mobilizing swiftly and effectively to prevent runaway climate change is heroically difficult. This view of reality has forced me again to accept that all efforts to prevent catastrophic levels of global warming may very well fail.

For me, responding to climate change in the right way has required me to fully surrender to the likelihood of failure. Before I gave up, I was always thinking about how the essay, personal action, or political activity I was working on was pointless because it was too small or too poorly executed to change the course of global warming. Understanding the tremendous human resistance to change helped me give up. Taking action could then become something worth doing not because success was likely but simply because the future is unknowable and because taking action is the right thing to do.

The Darwinian angle—adapt or go extinct

As the climate destabilizes, many things may begin to come apart. Depending on how quickly temperatures rise, it's likely that living things on Earth will experience radical changes in the coming decades. Soon, weather will probably be much more chaotic and unpredictable. There may be devastating crop failures, famines, mass human migrations, and global wars over water or food or land in habitable climates. Governments may collapse. The ocean food web may disintegrate. Species extinction rates may skyrocket. Chaos and suffering are possible, perhaps likely; yet it is impossible to predict exactly how, where, when, or on what scale.

The way I've begun to make sense of this angle is to accept that as a species, so far we've been unable to pull together and make the necessary changes to curtail our planetary climate impacts to those that human civilization can withstand in its current configuration. Perhaps we will dramatically change course very soon. Perhaps Earth's climate will stabilize at a level that can continue to sustain us. But if not, the only way forward may be global climate destabilization that either reroutes us into a species the planet can sustain or causes our extinction. Because I've now fully engaged in working to make it otherwise, I'm finding I can accept this possibility. As a biologist, I can appreciate that life on Earth—even human life—is ultimately subject to the rules of natural selection and evolution.

The capacity-building angle—a fresh perspective

I met forest ecologist Fred Swanson for lunch just before going off to the H. J. Andrews Experimental Forest to write. Swanson is tall and lean, with warm brown eyes, a full salt-and-pepper beard, and youthful energy that makes it difficult to guess his age. Swanson's research at the Andrews stretches back to the 1970s, and over lunch we talked about the forest, its history, and the work going on there now. I was feeling pretty down about the state of the world, my work, and pretty much everything else. I was disappointed and upset with President Obama's failure to make progress to combat climate change, and I was worried about Lia's future. However, during lunch Swanson said something that eventually grew into a fresh angle from which to envision global warming.

He told me that the Andrews Forest was the site of many of the original studies examining the ecology of old-growth forests and the biology of the northern spotted owl. Later, this research played a central role in the timber wars that erupted in Oregon, Washington, and northern California over old-growth logging and northern spotted owls. Swanson remembers when scientists were studying spotted owls before they were listed under the Endangered Species Act and before the battles over the fate of the last old-growth forests along the Pacific coast began. His experiences have affected his views on how today's critical conservation struggles might unfold.

When I told Swanson of my pessimism about the state of the world, the damage we're doing to the planet's living systems, and the threats we face, I noted his reaction. During this bleak phase of my life, when a conversation turned to climate change people generally responded in one of two ways: either with a quick flinch and an effort to change the subject or with a flash of recognition and a look of despair that matched my own. Swanson did not do either of these things. Instead he said, think of it this way: right now, what we are doing is important preparation. We're developing evidence, knowledge, potential approaches, and information. We're doing the science and the philosophy; we're laying the groundwork. When things shift and the window of opportunity opens, this framework we're working on will be there, in place and ready. What's going on right now—what you are helping with—is capacity building for an unknowable future.

Fred Swanson struck me as extremely bright, articulate, and thoughtful; he didn't deny or minimize how bleak things seemed at the moment, yet he seemed totally at ease with what he knew and what he was doing. Although I couldn't share his perspective or figure out at the time how someone so intelligent and well informed could be so sanguine, I found some comfort in hearing what he said. It wasn't until several years later that I found my own place in building capacity. Only then did I begin to feel the truth of this angle and begin to share some of his ease.

Choosing what to do

Taken together, all these angles on climate change have led me to see more clearly how to think and feel. And they've illuminated a way for me to act that finally feels right. Everything I've learned has reinforced my sense that we are alive during what is probably the most pivotal moment in modern human history. Right now is, almost certainly, our best real chance to take action on climate change. Accepting that this is the critical moment is clarifying.

As I've said, I respect changing personal behaviors—using less electricity, driving less, becoming a vegetarian or eating less meat, putting solar panels on the roof, and trying to be an example—and have taken these steps myself. Here's the thing, though: I believe that reducing your carbon footprint alone does not qualify as viable engagement. I believe it's necessary to engage with the political and economic systems that are locking us into ruining our climate and thus our life support systems. This feels to me like a middle path—it sits between thinking my actions can fix this and giving up. This middle path is where I (and, I submit, you too) need to go. Give up yet engage meaningfully.

In the fall of 2012, I started looking for community, state, and national actions to join. I wrote letters and participated in local climate forums and protests to fight coal export terminals in the Pacific Northwest. That winter, I went to several meetings of the local chapter of 350.org. Nothing felt quite like a fit, but I hoped something would click.

The more I learned, the more I was drawn to Citizens' Climate Lobby. I liked their sharp focus on building political will to stabilize the climate. I agreed with their emphasis on seeking nonpartisan solutions and focusing on positive messaging. I had learned through painful experience that talking with people about the doom of where our climate is headed fails miserably. Placing a fee on carbon emissions, as CCL proposed, made sense to me. I realized that my highest loyalty was to a stable climate, not a political party or a particular approach. This group shared that with me. I went to their website and joined their introductory call. I became convinced that CCL's proposal was the most viable, fair, passable, and effective solution I had seen. If it could be passed, it seemed to me that a steadily rising carbon fee could quickly begin to bring down greenhouse gas emissions, giving us a fighting chance to stabilize Earth's climate. The more I learned, the more convinced I became.

Most people I spoke with had never heard of CCL and there was no local chapter, so in the spring of 2013, I helped organize and launch the Corvallis chapter of

Citizens' Climate Lobby. Rarely have I done anything so difficult and uncomfortable or that I felt more unqualified for.

I am a classic introvert—I prefer being alone, with my family, or with one or two close friends. All the things that needed doing—contacting strangers, running meetings, organizing workshops, giving talks, doing outreach—were really hard for me. I wasn't good at them either, and I'd worry about each email I sent and each meeting I ran. I had imagined friends participating, but my husband, Mark, was the only person I knew who stayed on. Most people who came to the first meetings stopped coming, and I felt sure the dwindling attendance was my fault. I couldn't keep up on the emails and tasks without falling behind with other things. Some climate activists I spoke with were openly hostile to a carbon tax because it is based on using the power of the market to transition away from greenhouse gas emissions, and being opposed to the market, they were opposed to the carbon tax. Then at one meeting a near-stranger criticized me for having poor leadership skills, and Mark's calm presence was all that kept me from quitting. I told myself I'd give it three more months and then decide what to do.

In the winter, things began to change. I realized that if I were going to keep doing this, I would have to leave my ego behind. I had to decide that doing a so-so job was better than not doing it at all. In fact, I told myself, I might be the worst leader of a CCL chapter in the history of the organization, and that was okay, because I was trying. I also had to get some help. There was an extremely quiet but smart and reliable guy who had been there from the start, and he agreed to be co-leader of the group. I told myself that I couldn't control who came to meetings or what they said. I moved the meetings from my living room to a community meeting room because inviting people I barely know into my house was just too difficult. I began to accept that I wouldn't always keep up on CCL tasks, and that some emails simply wouldn't get answered. I worked to stop asking how well I was doing and just tried to be myself.

That is when I began to see that each of us has a place to be and a part to play. The trick is to ask what yours is and then do it. I began to understand that this was what I wanted to do, and that I could keep doing it. Monthly meetings became less difficult, and then sometimes they were actually fun. Our group began to have some successes. We met with our congressman, did outreach, and got published in our newspaper. I went to CCL conferences and met people who cared just as much as I do. I began giving talks, contacting congressional staffers and learning how to lobby. And one day it dawned on me that this group of CCL volunteers—many of them older male retirees that I'd initially felt so uncomfortable around—had somehow become my people.

When asked what I do, I used to avoid telling people that I write about climate change. I was sensitive to the discomfort, guilt, or awkwardness that often followed, so I usually just said I'm a writer. Once I started working with CCL, there was just no avoiding the fact that all my work is about combatting global warming. Recently, I ran into an acquaintance who asked me what I do, and without much thought I said I'm a climate writer and activist. In that moment, I realized I wasn't really concerned about her reaction, and it felt good to own my identity. I understand now that we live in extraordinary times that require us to find a way to speak up and to do something.

So, what did I learn? How did I get from worry, guilt, and despair to an unusual form of optimism, which feels like a kind of grace? Yesterday, our morning paper's biggest headline read, "Forest Chief: Tough Wildfires 'New Normal.'" It was, to me, a sign of how much things have changed since I began writing this book. We are, indeed, at the point where we can't ignore the changes anymore, and this is a good thing—in fact, it's brilliant. Headlines about global warming give me optimism. So the question is, what's next?

I want to leave you with a final paradox, illustrated by contradictory ideas about envisioning the future from two of the people I admire most in the world. One of those people is Wendell Berry, who said in an essay that "the future does not exist until it has become the past."[8] His point is that the future is unknowable, so predictions about it are pointless. Acting rightly in the present, he believes, is the best we can do. On the other hand, I once heard Mark Reynolds, the executive director of CCL and another hero of mine, say that the best predictor of how well at-risk youth will do is not how much they have endured in their past but instead what they imagine is possible in their future. His point, I think, is that to do well in our lives, we must be able to imagine—to believe in—a future we want.

Both men are absolutely right: the future is completely unknowable, yet how we imagine it profoundly affects us in our present. I learned I had to find a way to imagine a future I want to fight for and then act accordingly. Many of the possibilities before us are bleak and morally unjustifiable, so being a stand-up person requires taking action in spite of all the reasons not to. Taking action with spirit, with purpose, and, yes, even with optimism requires imagining that it is still possible for humanity to stabilize the climate and then finding a way to participate. That is ultimately what has set me free. I hope you find your way to imagining a good and desirable future, and then find your role in working to make it happen.

Take Action

Suggestions from the author about climate solution organizations to join or support:

350.org, *350.org*

Alliance for Climate Education, *acepace.org*

Amazon Conservation Association, *amazonconservation.org*

Carbonfund.org, *carbonfund.org*

Citizens' Climate Lobby/Citizens' Climate Education, *citizensclimatelobby.org*

Climate Solutions, *climatesolutions.org*

The Climate Reality Project, *climaterealityproject.org*

Environmental Defense Fund, climate initiatives, *edf.org*

The League of Conservation Voters, global warming work, *lcv.org*

More than Scientists, *morethanscientists.org*

Natural Resources Defense Council, climate change work, *nrdc.org*

Our Children's Trust, *ourchildrenstrust.org*

Rainforest Action Network, *ran.org*

Sierra Club, beyond fossil fuels programs, *sierraclub.org*

The Union of Concerned Scientists, global warming program, *ucsusa.org*

Voces Verdes, *vocesverdes.org*

Acknowledgements

Thank you to my amazing writing group, Lorraine Anderson, Carol Savonen, and Lee Sherman, who have read countless drafts, provided unending encouragement and support, and been true friends and wonderful companions throughout the writing of this book. You are all such stellar women, and I could not have finished this book without you.

Thank you to George Estreich, Jen Myers, Lael Leslie, Mike Weaver, and Monica Woelfel for miscellaneous help and encouragement. Thank you to Jim Baldwin for reading and editing an early draft—your excellent advice and editing skills helped immeasurably.

Thank you to Bill McKibben, Carly Lettero, Charles Goodrich, Fred Swanson, Kathleen Dean Moore, James Hansen, Madeleine Para, Mary Christina Wood, and Wendell Berry for inspiring me to do more and try harder. Thank you to Fred Strebeigh, Mark Bertness, Tom Ranker, Tom Kaye, and Yan Linhart for mentoring me and teaching me important things.

Thank you to Mark Reynolds, for helping me find my

path; I will be forever grateful. Thank you for your stories, your bottomless optimism, and your tears. Also, thank you for your emails—your love of exclamation points always lifts my spirits!

Thank you to all CCL supporters and climate activists everywhere for everything you do to work for a stable climate. Without you, this book would have ended very differently.

Thank you to Myra, Burt, Laura, and Karen Wise, Josh Gutwill, Jeannie Mayjor, Theresa Filtz, Dana Greci and Linda Hardison for your love, friendship, and kind patience with my climate obsession. Thank you to Sheridan McCarthy and Stanton Nelson at Meadowlark Publishing Services, for your hard work and patience helping me make the book look just the way I wanted it to; it was a real pleasure to work with you.

Extra special thank you to Lorraine Anderson, editor extraordinaire. You are, without a doubt, the best editor I have ever worked with. Your editorial skills, patience, kindness, excellent taste, attention to detail, and generosity improved the manuscript immensely without changing its fundamental character. I feel so lucky to have been able to work with you.

Finally, thank you to my beloved Mark and Lia, for your love and encouragement all along the way. Mark, thank you for believing I have something unique and valuable to say. Thank you for your patience as the process of writing this book dragged on, and for your willingness to be the one with the dependable paycheck. And to Lia, for filling the last 16 years of my life with your beautiful spirit and your kind, generous, wise self. I love you both infinity.

Notes

Starting Premises

1. National Research Council, *Climate Change: Evidence, Impacts, and Choices* (National Academy of Sciences, 2012), p. 2; J. Walsh et al., "Our Changing Climate," Chapter 2 in *Climate Change Impacts in the United States: The Third National Climate Assessment* (U.S. Global Change Research Program, 2014), pp. 54–55. This latter piece of writing is the compelling, understandable, brief yet accurate explanation of human-caused climate change for nonscientists that I was looking for and never found before I started writing this book.

2. Intergovernmental Panel on Climate Change, "Summary for Policymakers," in *Climate Change 2013: The Physical Science Basis,* Contribution of Working Group I to the Fifth Assessment Report of the Intergovernmental Panel on Climate Change (Cambridge University Press, 2013).

3. Agreeing on the exact amount of warming requires choosing a year to start (1880 in this case). The 0.85 degrees C is from the National Research Council's 2012 report *Climate Change: Evidence, Impacts, and Choices.*

4. Matt McGrath, "Warming Set to Breach 1C Threshold," *BBC News,* November 9, 2015; James Hansen et al., "Global Temperature in 2015," January 19, 2016.

5. National Research Council, *Climate Change.*

6. For example, see Trevor M. Letcher (ed.), *Climate Change: Observed Impacts on Planet Earth* (Elsevier, 2009); Gavin Schmidt and Joshua Wolfe, *Climate Change: Picturing the Science* (Norton, 2009); and David Archer and Stefan Rahmstorf,

The Climate Crisis: An Introductory Guide to Climate Change (Cambridge University Press, 2010).

7. IPCC, *Climate Change 2013.*

8. Kevin E. Trenberth, "Has There Been a Hiatus?" *Science* 349 (2015): 691–92; Justin Gillis, "Global Warming 'Hiatus' Challenged by NOAA Research," *New York Times,* June 4, 2015.

9. National Research Council, *Climate Change,* 15–17.

10. Camille Parmesan, "Ecological and Evolutionary Responses to Recent Climate Change," *Ann Rev. Ecol. Syst.* 37 (2006): 637–69; and I-Ching Chen et al., "Rapid Range Shifts of Species Associated with High Levels of Climate Warming," *Science* 333 (2011): 1024–26.

11. National Research Council, *Climate Change,* 17.

12. J. Zalasiewicz et al., "The New World of the Anthropocene," *Environmental Science and Technology* 44 (2010): 2228–31; Matt McGrath, "Climate Impacts 'Overwhelming' — UN," *BBC News,* March 31, 2014.

13. National Snow and Ice Data Center, "2015 Melt Season in Review," Arctic Sea Ice News and Analysis, October 6, 2015.

14. National Research Council, *Climate Change,* 15–17.

15. Julia Whitty, "The End of a Myth," *OnEarth* 34 (Spring 2012): 39–43.

16. William R. L. Anderegg et al., "Consequences of Widespread Tree Mortality Triggered by Drought and Temperature Stress," *Nature Climate Change* 3 (2013): 30–36; Chris Mooney, "The Forests of the World Are in Serious Trouble, Scientists Report," *Washington Post,* August 20, 2015.

17. Justin Gillis, "A Warming Planet Struggles to Feed Itself," *New York Times,* June 4, 2011; J. Hansen, M. Sato, and R. Ruedy, "Perception of Climate Change," *Proc. Nat. Acad. Sci.* 109 (2012): E2415–E2423.

18. Gillis, "A Warming Planet"; W. Schlenker and M. J. Roberts, "Nonlinear Temperature Effects Indicate Severe Damages to U.S. Crop Yields Under Climate Change," *Proc. Nat. Acad. Sci.* 106 (2009): 15594–98; T. Wheeler and J. von Braun, "Climate Change Impacts on Global Food Security," *Science* 341 (2013): 508–13.

19. H. Riebeek, "Global Warming," NASA Earth Observatory website (earthobservatory.nasa.gov), June 3, 2010.

20. S. Rahmstorf and D. Coumou, "Increase of Extreme Events in a Warming World," *Proc. Nat. Acad. Sci.* 108 (2011): 17905–09; Intergovernmental Panel on Climate Change,

"Summary for Policy Makers," in *Managing the Risks of Extreme Events and Disasters to Advance Climate Change Adaptation,* a special report of Working Groups I and II of the Intergovernmental Panel on Climate Change (Cambridge University Press, 2012), pp. 1-19.

21. Jim Algar, "Antarctic Glaciers' Melting Said Near 'Unstoppable' Point, Threatening Sea Level Rise," *Tech Times,* November 4, 2015.

22. More details about this graph and the four RCPs can be found in IPCC, *Climate Change 2013,* Box SPM.1 and Chapter 12.

23. Riebeek, "Global Warming."

24. In IPCC language, the term *likely* here indicates with 66 to 100 percent probability.

25. Richard A. Kerr, "The IPCC Gains Confidence in Key Forecast," *Science* 342 (2013): 23–24.

26. Ivy Tan et al., "Observational Constraints on Mixed-phase Clouds Imply Higher Climate Sensitivity," *Science* 352 (2016): 224-227.

27. Figures, tables, and text throughout IPCC, *Climate Change 2013,* show detailed projections that back up this statement; see in particular Chapter 12, "Long-Term Climate Change: Projections, Commitments and Irreversibility," and TFE.5, "Irreversibility and Abrupt Change."

28. Malte Meinshausen et al., "Greenhouse-Gas Emission Targets for Limiting Global Warming to 2°C," *Nature* 458 (2009): 1158–62.

29. Justin Gillis, "U. N. Climate Panel Endorses Ceiling on Global Emissions," *New York Times,* September 27, 2013; Coral Davenport et al., "Inside the Paris Climate Deal," *New York Times,* December 12, 2015.

30. Davenport et al., "Inside the Paris Climate Deal."

31. James Hansen et al., "Assessing 'Dangerous Climate Change,': Required Reduction of Carbon Emissions to Protect Young People, Future Generations, and Nature," *PLOS ONE* 8 (2013): 1–26.

32. P. Friedlingstein et al., "Persistent Growth of CO2 Emissions and Implications for Reaching Climate Targets," *Nature Geoscience* 7 (2014): 709–15.

33. Bobby Magill, "Energy Bombshell: CO2 Emissions Stabilized in 2014," Climate Central, March 13, 2015, and Justin Gillis and Chris Buckley, "Period of Soaring Emissions May

Be Ending, New Data Suggest," *New York Times,* December 7, 2015.

34. Laura Barron-Lopez, "Report: Globe Could Pass UN Warming Target," *The Hill,* September 8, 2014; IPCC, *Climate Change 2013.*

35. Justin Gillis, "Climate Maverick to Quit NASA," *New York Times,* April 1, 2013; Scott Learn, "Global Warming Activist, Former NASA Scientist James Hansen Speaks Out Before World Affairs Council of Oregon Appearance," *The Oregonian,* April 24, 2013; Mark Drajem, "NASA's Hansen Arrested Outside White House at Pipeline Protest," *Bloomberg,* August 29, 2011.

36. International Energy Agency, "Executive Summary," *World Energy Outlook 2011,* November 2011, pp. 39–45.

37. Richard Kerr, "Bleak Prospects for Avoiding Dangerous Global Warming," *Science Now,* October 23, 2011; Ben Garside, "World Falls Behind in Efforts to Tackle Climate Change: Report," *Scientific American,* September 7, 2014; Marianne Lavelle, "Climate Report 'Introduces Sobriety' to Paris," *Science Insider,* June 15, 2015.

38. James Hansen, "Global Warming: Is There Still Time to Avoid Disastrous Human-Made Climate Change? i.e. Have We Passed a 'Tipping Point'?" Discussion on April 23, 2006, at the National Academy of Sciences, Washington, DC; James Hansen, "How Can We Avert Dangerous Climate Change?" Revised and expanded from written testimony on Energy Independence and Global Warming, U.S. House of Representatives, April 26, 2007.

39. J. Hansen et al., "Target Atmospheric CO2: Where Should Humanity Aim?" *Open Atmos. Sci J.* 2 (2008): 217–31; Hansen, "Assessing 'Dangerous Climate Change'."

40. American Association for the Advancement of Science, "What We Know," whatweknow.aaas.org, 2014; A. Morales, "Damage from Global Warming Soon Will Be Irreversible, Says Leaked UN Report," *National Post,* August 26, 2014; National Research Council, *Climate Change.*

Chapter 1. Extreme Weather

1. National Oceanic and Atmospheric Administration, "Extreme Weather 2011," NOAA website, 2012.

2. Intergovernmental Panel on Climate Change, *Managing the Risks of Extreme Events and Disasters to Advance Climate Change*

Adaptation, a special report of Working Groups I and II of the Intergovernmental Panel on Climate Change (Cambridge University Press, 2012).

3. National Climatic Data Center, "NCDC Releases 2012 Billion-dollar Weather and Climate Disasters Information," NOAA website, 2013.

4. Amanda Paulson, "For Colorado's 'Biblical' Floods, Numbers Tell Astonishing Tale (+Video)," *Christian Science Monitor,* September 18, 2013.

5. National Climatic Data Center, "Billion-Dollar Weather and Climate Disasters: Table of Events," National Centers for Environmental Information, NOAA website, 2014.

6. Lydia O'Connor, "2015 Has Been a Year of Record-Breaking U.S. Weather Events," *Huffington Post,* October 5, 2015; Katherine Bagley, "2015: The Year the Weather Took a Particularly Wild Ride," *InsideClimate News,* December 23, 2015.

7. Ian Simpson, "U.S. Wildfires Burned Record Area in 2015: Agriculture Department," Reuters website, January 6, 2016.

8. "96 percent of Americans Live in Counties Recently Hit by Weather Disaster," Environment America News Release, November 12, 2015.

9. Christopher Joyce, "Science Confirms 2014 Was Hottest Yet Recorded, on Land and Sea," National Public Radio, July 17, 2015; Justin Gillis, "2015 Was Hottest Year in Historical Record, Scientists Say," *New York Times,* January 20, 2016.

10. Alyson Kenward, "What's With the Weather? Is Climate Change to Blame?" *Yale Environment 360,* April 21, 2011.

11. John Metcalfe, "NASA's Alarming Map of the Worst Australian Heat Wave on Record," *The Atlantic Citylab,* January 24, 2013.

12. Quirin Schiermeier, "Did Climate Change Cause Typhoon Haiyan?" *Nature,* November 11, 2013; D. Normile, "Clues to Supertyphoon's Ferocity Found in the Western Pacific," *Science* 342 (2013): 1027.

13. Tom Miles, "Weather Disasters Occur Almost Daily," Reuters website, November 23, 2015.

14. "Heat Waves: The Details," in "Heat Waves and Climate Change: A Science Update from Climate Communication—June 28, 2012," Climate Communication website; Brian Kahn, "Climate Change Is Increasing Extreme Heat Globally," Climate Central website, February 26, 2014.

15. Hansen et al., "Perception of Climate Change," *Proc. Natl. Acad. Sci.* 109 (2012): 14726–27.

16. "Heat Waves: The Details," Climate Communication website.

17. "Precipitation, Floods and Drought," Climate Communication website.

18. Ibid.

19. A. Dai, "Drought Under Global Warming: A Review," *Wiley Interdisciplinary Reviews: Climate Change* 2 (2011): 45–65.

20. "Precipitation, Floods and Drought" and "Drought," Climate Communication website.

21. Schiermeier, "Did Climate Change Cause Typhoon Haiyan?"

22. "Hurricanes and Other Summer Storms," Climate Communication website.

23. Justin Gillis, "Harsh Political Reality Slows Climate Studies Despite Extreme Year," *New York Times,* December 24, 2011.

24. Andrew R. Solow, "Extreme Weather, Made By Us?" *Science* 349 (2015): 1444–45.

25. P. A. Stott et al., "Human Contribution to the European Heatwave of 2003," *Nature* 432 (2004): 610–14.

26. John Carey, "Global Warming and the Science of Extreme Weather," *Scientific American,* June 29, 2011.

27. Nathaniel Gronewold, "Is the Flooding in Pakistan a Climate Change Disaster?" *Scientific American,* August 18, 2010.

28. Hansen et al., "Perception."

29. "New Analyses Find Evidence of Human-Caused Climate Change in Half of the 12 Extreme Weather and Climate Events Analyzed From 2012," NOAA website, September 5, 2013.

30. Brian Kahn, "A Timeline of 2013 Extreme Weather and Global Warming," Climate Central website, September 29, 2014.

31. John Schwartz, "Scientists Study Links Between Climate Change and Extreme Weather," *New York Times,* November 5, 2015.

32. Seth Borenstein, "Study Blames Global Warming for 75 Percent of Very Hot Days," Associated Press website, April 27, 2015.

33. Kenward, "What's with the Weather?"

Chapter 2. Life on Land

1. Darcy Frey, "George Divoky's Planet," *New York Times Magazine,* January 6, 2002.

2. Camille Parmesan, "Ecological and Evolutionary Responses to Recent Climate Change," *Ann. Rev. Ecol. Syst.* 37 (2006): 637–69; I-Ching Chen et al., "Rapid Range Shifts of Species Associated with High Levels of Climate Warming," *Science* 333 (2011): 1024–26.

3. Camille Parmesan and G. Yohe, "A Globally Coherent Fingerprint of Climate Change Impacts Across Natural Systems," *Nature* 421 (2003): 37–42.

4. Chen, "Rapid Range Shifts."

5. Justin Gillis, "Spared Winter Freeze, Florida's Mangroves Are Marching North," *New York Times,* December 30, 2013.

6. Parmesan, "Ecological and Evolutionary Responses."

7. D. W. Inouye et al., "Climate Change Is Affecting Altitudinal Migrants and Hibernating Species," *Proc. Natl. Acad. Sci.* 97 (2000): 1630–33.

8. Parmesan, "Ecological and Evolutionary Responses," 645.

9. Ibid., 647.

10. Katie Valentine, "Penguin Chicks in Argentina Are Dying from Increased Heavy Rains and Extreme Heat," *Think Progress,* January 30, 2014.

11. William Laurance, "The World's Tropical Forests Are Already Feeling the Heat," *Yale Environment 360,* May 2, 2011.

12. Justin Welbergen et al., "Killer Climate: Tens of Thousands of Flying Foxes Dead in a Day," The Conversation, February 24, 2014.

13. Justin Welbergen, email message to the author, January 28, 2012.

14. Jim Robbins, "What's Killing the Great Forests of the American West?" *Yale Environment 360,* March 15, 2010; Justin Gillis, "With Deaths of Forests, a Loss of Key Climate Protectors," *New York Times,* October 1, 2011.

15. Joe Romm, "Global Warming Is Doubling Bark Beetle Mating, Boosting Tree Attacks Up To 60-fold, Study Finds," *Think Progress,* April 30, 2012.

16. Jane Kay, "Warming Temperatures Help Endangered Swan Rebound," *The Daily Climate,* January 22, 2012.

17. Parmesan, "Ecological and Evolutionary Responses," 646.

18. Scott C. Doney et al., "Climate Change Impacts on Marine Ecosystems," *Annu. Rev. Marine Sci.* 4 (2012): 11–37.

19. Andy Isaacson, "In a Changing Antarctica, Some Penguins Thrive as Others Suffer," *New York Times,* May 9, 2011.

20. J. A. Pounds et al., "Widespread Amphibian Extinctions from Endemic Disease Driven by Global Warming," *Nature* 439 (2006): 161–67; J. A. Pounds and L. A. Coloma, "Beware the Lone Killer," *Nature Reports Climate Change* 2 (2008): 57–59.

21. Parmesan, "Ecological and Evolutionary Responses"; C. D. Thomas et al., "Range Retractions and Extinction in the Face of Climate Warming," *TREE* 21 (2006): 415–16.

22. Cally Carswell, "Bumblebees Aren't Keeping Up with a Warming Planet," *Science* 349 (2015): 126–27.

23. Katherine Bagley, "Climate Change Spells Extinction for a Mountain Species, Research Shows," *InsideClimate News,* April 28, 2011.

24. Ed Struzik, "Northern Mystery: Why Are Birds of the Arctic in Decline?" *Yale Environment 360,* January 23, 2014.

25. Jan Zalasiewicz et al., "The New World of the Anthropocene," *Environmental Science and Technology* 44 (2010): 2228–31.

26. Ibid.

27. Carl Zimmer, "Study Finds Climate Change as Threat to 1 in 6 Species," *New York Times,* April 30, 2015.

28. "Monitoring Climate Change with Arctic Sea Birds," Friends of Cooper Island website; "Dr. George Divoky, Seabird Biologist," NSF Arctic Stories website.

Chapter 3. Oceans

1. Julia Whitty, "The End of a Myth," *OnEarth* 34 (Spring 2012): 39–43.

2. "Climate Change and the Oceans," New England Aquarium website, 2015.

3. Whitty, "The End," 40.

4. "What Is the Global Ocean Conveyer Belt?" NOAA National Ocean Service website.

5. "The Great Ocean Conveyer Belt," Environmental Literacy Council website, April 8, 2008.

6. John Abraham, "The Oceans Are Warming So Fast, They Keep Breaking Scientists' Charts," *The Guardian,* January 22, 2015.

7. Scott C. Doney et al., "Climate Change Impacts on Marine Ecosystems," *Annu. Rev. Marine Sci.* 4 (2012): 11–37.

8. Henry Fountain, "A Gulf in Ocean Knowledge," *New York Times,* October 6, 2014.

9. Craig Welch, "Sea Change: Pacific Ocean Takes Perilous Turn," *Seattle Times,* September 12, 2013.

10. Doney et al., "Climate Change Impacts."

11. Ibid.

12. "State of the Cryosphere: Contribution of the Cryosphere to Changes in Sea Level," National Snow and Ice Data Center website, July 13, 2015.

13. E. S. Poloczanska et al., "Global Imprint of Climate Change on Marine Life," *Nature Climate Change* 3 (2013): 919–25.

14. Doney et al., "Climate Change Impacts."

15. Ibid.

16. "Climate Change Effects on Ocean Animals," New England Aquarium website.

17. Camille Parmesan, "Ecological and Evolutionary Responses to Recent Climate Change," *Ann. Rev. Ecol. Syst.* 37 (2006): 637–69.

18. Tom Philpott, "Something Really, Really Terrible Is About to Happen to Our Coral," *Mother Jones,* January 28, 2015.

19. Doney et al., "Climate Change Impacts."

20. "NOAA Declares Third Ever Global Coral Bleaching Event," NOAA website, October 8, 2015.

21. Elizabeth Kolbert, "Our Darkening Sea: What Carbon Emissions Are Doing to the Ocean," in Bill McKibben (ed.), *The Global Warming Reader* (Penguin Books, 2012), 377–98.

22. Doney et al., "Climate Change Impacts."

23. Ibid.

24. Michelle Innis, "Climate-Related Death of Coral Around World Alarms Scientists," *New York Times,* April 9, 2016.

25. Kolbert, "Our Darkening Sea," 384–85.

26. "Climate Change and the Oceans," New England Aquarium website.

27. Pat Brennan, "NASA: Polar Ice Sheets Are Melting," *Orange County Register,* December 5, 2012.

28. Jim Algar, "Antarctic Glaciers' Melting Said Near 'Unstoppable' Point, Threatening Sea Level Rise," Tech Times, November 4, 2015.

29. "All About Sea Ice," National Snow and Ice Data Center, 2015.

30. Figure 3, "2015 Melt Season in Review," Arctic Sea Ice News and Analysis, National Snow and Ice Data Center, October 6, 2015.

31. "State of the Cryosphere: Sea Ice," National Snow and Ice Data Center, November 2, 2015.

32. J. Walsh et al., "Our Changing Climate," in *Climate Change Impacts in the United States: The Third National Climate Assessment,* J. M. Melillo et al. (eds.), U.S. Global Change Research Program, 2014, pp. 54–55.

33. Doney et al., "Climate Change Impacts."

34. "Climate Change and the Oceans," New England Aquarium website.

35. "All About Sea Ice," National Snow and Ice Data Center.

36. Intergovernmental Panel on Climate Change, "Summary for Policymakers," in *Climate Change 2013: The Physical Science Basis,* 25–26.

37. Ben Strauss et al., "Surging Seas," a Climate Central Report, March 14, 2012.

38. Walsh, "Our Changing Climate."

39. IPCC, "Summary."

40. Walsh, "Our Changing Climate."

41. Ibid.

42. Strauss et al., "Surging Seas."

43. Ibid.

44. Justin Gillis, "The Flood Next Time," *New York Times,* January 13, 2014.

45. Brian Fagan, "As Extreme Weather Increases, Bangladesh Braces for the Worst," *Yale Environment 360,* June 6, 2013.

46. Algar, "Antarctic Glaciers' Melting."

47. "The Global Conveyer Belt," NOAA National Ocean Service website, July 10, 2013.

48. "The Great Ocean Conveyer Belt," Environmental Literacy Council website, August 1, 2015.

49. Chris Mooney, "Global Warming Is Now Slowing Down the Circulation of the Oceans—With Potentially Disastrous Consequences," *Washington Post,* March 23, 2015.

50. "Climate Change and the Oceans," New England Aquarium.

51. Mooney, "Global Warming Is Now Slowing."

52. Whitty, "The End."

Chapter 4. Forests and Trees

1. Jim Robbins, "Why Trees Matter," *New York Times* Opinion, April 11, 2012.

2. Justin Gillis, "A Scientist, His Work and a Climate Reckoning," *New York Times,* December 21, 2010.

3. Yude Pan et al., "A Large and Persistent Carbon Sink in the World's Forests," *Science* 333 (2011): 988–93.

4. Gordon B. Bonan, "Forests and Climate Change: Forcings, Feedbacks, and the Climate Benefits of Forests," *Science* 320 (2008): 1444–1449.

5. Robert Krier, "Trees Absorb Less Carbon in Warming World Than Experts Have Assumed," *InsideClimate News,* May 23, 2012.

6. "The Forest Biome," University of California Museum of Paleontology.

7. Pan et al., "Large and Persistent Carbon Sink."

8. William R. L. Anderegg et al., "Consequences of Widespread Tree Mortality Triggered by Drought and Temperature Stress," *Nature Climate Change* 3 (2013): 30–36.

9. Chris Mooney, "The Forests of the World Are in Serious Trouble, Scientists Report," *Washington Post,* August 20, 2015.

10. University of Leeds, "Two Severe Amazon Droughts in Five Years Alarms Scientists," *ScienceDaily* press release, February 3, 2011; Robbins, "Why Trees Matter."

11. Justin Gillis, "Amazon Forest Becoming Less of a Climate Change Safety Net," *New York Times,* March 23, 2015.

12. Justin Gillis, "With Deaths of Forests, a Loss of Key Climate Protectors," *New York Times,* October 1, 2011.

13. Jim Robbins, "The Rapid and Startling Decline of World's Vast Boreal Forests," *Yale Environment 360,* October 12, 2015.

14. P. J. van Mantgem et al., "Widespread Increase of Tree Mortality Rates in the Western United States," *Science* 323 (2009): 521–34.

15. Gillis, "With Deaths of Forests"; Justin Gillis, "Climate Change Threatens to Kill Off More Aspen Forests by 2050s, Scientists Say," *New York Times,* March 30, 2015.

16. Robbins, "Why Trees Matter."

17. Lisa Hayward, "Pining Away: Climate Change, Mountain Pine Beetles, and the Future of Whitebark Pine," *Northwest Climate Magazine,* May 2015, 14–16.

18. Jim Robbins, "What's Killing the Great Forests of the American West?" *Yale Environment 360,* March 15, 2010; Gillis, "With Deaths of Forests."

19. Robbins, "What's Killing the Great Forests?"

20. A. L. Westerling et al., "Warming and Earlier Spring Increase Western U.S. Forest Wildfire Activity," *Science* 313 (2006): 940–43.

21. Bob Berwyn, "Link Between Global Warming and Wildfires Becoming More Clear," Summit County Citizens Voice website, October 3, 2012; "The Age of Western Wildfires," Climate Central, September 18, 2012.

22. Maret Pajutee, "After the Pole Creek Fire: Fire Effects and Lessons from 10 Years of Big Fires in Sisters," Deschutes Land Trust website, January 2013.

23. National Research Council, *Climate Stabilization Targets: Emissions, Concentrations, and Impacts Over Decades to Millennia* (Washington, DC: National Academies Press, 2011).

24. Trevor Hughes, "Climate Change Accelerating Death of Western Forests," *USA Today,* September 10, 2014; John Schwartz, "As Fires Grow, a New Landscape Appears in the West," *New York Times,* September 21, 2015.

25. U.S. Forest Service, "The Rising Cost of Wildfire Operations: Effects on the Forest Service's Non-Fire Work," USDA Report, August 4, 2015.

26. Ed Struzik, "A New Global Tinderbox: The World's Northern Forests," *Yale Environment 360,* October 1, 2015.

27. Jim Robbins, "What's Killing the Great Forests?"

28. Scott L. Stephens et al., "Temperate and Boreal Mega-Fires: Characteristics and Challenges," *Front. Ecol. Environ.* 12 (2014): 115–22.

29. Gillis, "With Deaths of Forests."

30. John Carey, "A Scientist Extols the Value of Forests Shaped by Humans," *Yale Environment 360* interview with Susanna Hecht, April 5, 2011; F. Bongers et al., "The Potential of Secondary Forests," *Science* 348 (2015): 642–43.

31. Z. Gedalof and A. A. Berg, "Tree Ring Evidence for Limited Direct CO2 Fertilization of Forests over the 20th Century," *Global Biogeochemical Cycles* 24 (September 2010).

32. Gillis, "With Deaths of Forests."

33. Pan et al., "Large and Persistent Carbon Sink."

34. Anderegg et al., "Consequences of Widespread Tree Mortality"; Michael D. Lemonick, "Climate Change Stress Killing Forests, and Why It Matters," *Climate Central*, September 9, 2012.

35. Craig D. Allen et al., "On Underestimation of Global Vulnerability to Tree Mortality and Forest Die-off from Hotter Drought in the Anthropocene," *Ecosphere* 6 (2015): 1–55.

36. Robbins, "Why Trees Matter."

Chapter 5. Agriculture

1. D. B. Lobell et al., "Climate Trends and Global Crop Production since 1980," *Science* 333 (2011): 616–20.

2. Justin Gillis, "A Warming Planet Struggles to Feed Itself," *New York Times*, June 4, 2011.

3. Sophie Wenzlau, "Global Food Prices Continue to Rise," Vital Signs, Worldwatch Institute website, April 11, 2013; Getaw Tadesse et al.,"Drivers and Triggers of International Food Price Spikes and Volatility," Food Policy 47 (2014): 117–28.

4. S. P. Long et al., "Food for Thought: Lower-than-expected Crop Yield Simulation with Rising CO2 Concentrations," *Science* 312 (2006): 1918–21; Earth Institute at Columbia University, "Toll of Climate Change on World Food Supply Could be Worse than Thought," ScienceDaily, December 4, 2007.

5. Brian K. Sullivan et al., "Extreme Weather Wreaking Havoc on Food as Farmers Suffer," *Bloomberg Business*, January 17, 2014; Nicholas St. Fleur, "Drought and Heat Took a Heavy Toll on Crops, Study Finds," *New York Times*, January 6, 2016.

6. W. Schlenker and M. J. Roberts, "Nonlinear Temperature Effects Indicate Severe Damages to U.S. Crop

Yields Under Climate Change," *Proc. Nat. Acad. Sci.* 106 (2009): 15594–98.

7. Gillis, "A Warming Planet."

8. J. Schmidhuber and F. Tubiello, "Global Food Security Under Climate Change," *Proc. Nat. Acad. Sci.* 104 (2007): 19703–08; Jeanne Roberts, "Climate Change Will Challenge Farmers as Crop Pests Spread," InsideClimate News, August 3, 2009.

9. "Climate Change's Effect on Crop Yields Worse Than Thought: Study," Australian Broadcasting Company website, March 16, 2014.

10. Joe Romm, "NASA Bombshell: Global Groundwater Crisis Threatens Our Food Supplies and Our Security," ClimateProgress website, October 31, 2014.

11. Justin Gillis, "Panel's Warning on Climate Risk: Worst Is Yet to Come," *New York Times,* March 31, 2014; Working Group II, Intergovernmental Panel on Climate Change, "Summary for Policymakers," *Climate Change 2014: Impacts, Adaptation, and Vulnerability,* March 31, 2014.

12. Schmidhuber and Tubiello, "Global Food Security"; D. B. Lobell et al., "Prioritizing Climate Change Adaptation Needs for Food Security in 2030," *Science* 319 (2008): 607–10.

13. Lobell et al., "Prioritizing"; D. Fogarty, "Global Study Reveals Climate Peril for Millions of Poor," Reuters website, June 2, 2011.

14. Wenzlau, "Global Food Prices"; Tadesse et al.,"Drivers and Triggers."

15. Lobell et al., "Prioritizing"; Gerald C. Nelson et al., "Climate Change: Impact on Agriculture and Costs of Adaptation," International Food Policy Research Institute Report, November 6, 2009; A. J. Challinor et al., "A Meta-analysis of Crop Yield Under Climate Change and Adaptation," *Nature Climate Change* 4 (2014): 287–91.

16. S. J. Vermeulen et al., "Climate Change and Food Systems," *Annu. Rev. Environ. Resour.* 37 (2012): 195–222; Paul C. West et al., "Leverage Points for Improving Global Food Security and the Environment," *Science* 345 (2014), 325–28.

17. J. P. Reganold et al., "Transforming U.S. Agriculture," *Science* 332 (2011): 670–71; Food and Agriculture Organization of the United Nations, *Climate-Smart Agriculture Sourcebook,* June 5, 2013.

18. P. J. Gerber et al., *Tackling Climate Change through Livestock – A Global Assessment of Emissions and Mitigation Opportunities* (Food and Agriculture Organization of the United Nations, 2013).

19. Gidon Eshel et al., "Land, Irrigation Water, Greenhouse Gas, and Reactive Nitrogen Burdens of Meat, Eggs, and Dairy Production in the United States," *Proc. Nat. Acad. Sci.* 111 (2014): 11996–12001.

20. Sara J. Scherr and Sajal Sthapit, "Mitigating Climate Change Through Food and Land Use," Worldwatch Report 179, 2009.

21. Gerber et al., *Tackling Climate Change through Livestock.*

22. S. J. Vermeulen et al., "Climate Change and Food Systems."

23. Scherr and Sthapit, "Mitigating Climate Change," and Gerber et al., *Tackling Climate Change through Livestock.*

24. Eshel et al., "Land, Irrigation Water"; West et al., "Leverage Points."

25. For example, see Gerber et al., *Tackling Climate Change through Livestock;* Vermeulen et al., "Climate Change and Food Systems"; West et al., "Leverage Points"; Scherr and Sthapit, "Mitigating Climate Change."

26. Robert Sanders, "Fertilizer Use Responsible for Increase in Nitrous Oxide in Atmosphere," UC Berkeley News Center press release, April 2, 1012.

27. Scherr and Sthapit, "Mitigating Climate Change."

28. Reganold et al., "Transforming U.S. Agriculture."

29. Scherr and Sthapit, "Mitigating Climate Change"; Justin Gillis, "Damaging the Earth to Feed Its People," *New York Times,* June 4, 2011.

30. D. Donlon, "The Agricultural Fulcrum: Better Food, Better Climate," *The Atlantic,* January 18, 2013; Eshel et al., "Land, Irrigation Water"; Rodale Institute, "How Organic Farming Can Reverse Climate Change," Ecowatch.com, April 22, 2014.

31. Joseph Mercola, "Bill Gates: One of the World's Most Destructive Do-gooders?" Mercola.com, March 4, 2012.

32. Glen Ashton, "Bill Gates' Support of GM Crops Is Wrong Approach for Africa," *Seattle Times,* Opinion, February 27, 2012.

33. Bill McKibben, *Eaarth* (St. Martin's Griffin, 2010).

Chapter 6. Paradoxes

1. Eric Wagner, "It's the End of the World As We Know It ... and I Feel Fine," *Conservation* 11 (2011): 48.

2. Paul Gilding, *The Great Disruption: How the Climate Crisis Will Transform the Global Economy* (Bloomsbury Press, 2011), p. 2.

3. This wording is borrowed from the Citizens' Climate Lobby website and is based on evidence from multiple reports, including W.R.L. Anderegg et al., "Expert Credibility in Climate Change," *Proc. Nat. Acad. Sci.* 107 (2010): 12107–9; and J. Cook et al., "Quantifying the Consensus on Anthropogenic Global Warming in the Scientific Literature," *Env. Res. Lett.* 8 (2013): 024024.

4. Seth Borenstein, "Climate Change as Certain as Cancer from Smoking, Scientists Say," *HeraldNet,* September 24, 2013.

5. Christopher F. Schuetze, "Scientists Agree Overwhelmingly on Global Warming. Why Doesn't the Public Know That?" *New York Times,* May 16, 2013.

6. Cary Funk and Lee Rainie, "Public and Scientists' Views on Science and Society," Pew Research Center, January 29, 2015.

Chapter 7. Blame and Moving On

1. Intergovernmental Panel on Climate Change, "Summary for Policymakers," in *Climate Change 2013: The Physical Science Basis,* Contribution of Working Group I to the Fifth Assessment Report of the Intergovernmental Panel on Climate Change (Cambridge University Press, 2013).

2. P. Friedlingstein et al., "Persistent Growth of CO2 Emissions and Implications for Reaching Climate Targets," *Nature Geoscience* 7 (2014): 709–15.

3. Final Primary Night Presumptive Democratic Nominee Speech, St. Paul, Minnesota, June 3, 2008, Obamaspeeches.com.

4. Bill McKibben,"Duty Dodgers," *Orion,* July/August 2010.

5. Bill McKibben, "The Attack on Climate-change Science," *The Nation,* February 25, 2010.

6. Neela Banerjee et al., "Exxon: The Road Not Taken," *InsideClimate News,* September 16, 2015; Naomi Oreskes, "Exxon's Climate Concealment," *New York Times,* Opinion, October 9, 2015.

7. Timothy Egan, "Can the Kochs Hold Back History?" *New York Times,* Opinion, May 8, 2014; "Koch Industries: Secretly Funding the Climate Denial Machine," Greenpeace website, Global Warming page.

8. Katherine Bagley, "Climate Scientists Face Organized Harassment in U.S.," Bloomberg.com, September 10, 2012.

9. Michael Mann, "The Assault on Climate Science," *New York Times,* Op-Ed, December 8, 2015.

10. For one critique, see Joe Romm, "Silence of the Lambs: Media Herd's Coverage of Climate Change 'Fell Off the Map' in 2010," ClimateProgress website, January 3, 2011.

11. Jody Warrick "Why Are So Many Americans Skeptical about Climate Change? A Study Offers a Surprising Answer," *Washington Post,* November 23, 2015.

12. Justin Gillis and John Schwartz, "Deeper Ties to Corporate Cash for Doubtful Climate Researcher," *New York Times,* February 21, 2015.

13. Victoria Bekiempis, "How the Media Fails to Cover Climate Change," *Newsweek,* June 26, 2014.

14. "Republicans, Clean Energy, and Climate Change," survey conducted for ClearPath, August 24–27, 2015; Anthony Leiserowitz et al., *Climate Change in the American Mind,* October 2015, Yale Project on Climate Change Communication and George Mason University Center for Climate Change Communication.

15. R. Kerr, "Amid Worrisome Signs of Warming, Climate Fatigue Sets In," *Science* 326 (2009): 926–28.

16. Denise Robbins, "This New Study Shows How The Media Makes People Climate Change Cynics—And What They Can Do Differently," Media Matters website, September 18, 2015; John Upton, "Media Contributing to 'Hope Gap' on Climate Change," Climate Central website, March 28, 2015.

17. George Marshall, *Don't Even Think About It: Why Our Brains Are Wired to Ignore Climate Change* (Bloomsbury Publishing, 2014); Kari Norgaard, *Living in Denial: Climate Change, Emotions, and Everyday Life* (MIT Press, 2011); Per Espen Stoknes, *What We Think About When We Try Not to Think About Global Warming: Toward a New Psychology of Climate Action* (Chelsea Green, 2015).

18. Bob Simison, "New York Attorney General Subpoenas Exxon on Climate Research," *InsideClimate News,* November 5, 2015.

19. Robert J. Brulle, "Institutionalizing Delay: Foundation Funding and the Creation of U.S. Climate Change Counter-Movement Organizations," *Climatic Change* 122 (2014): 681–94; Brendan DeMelle, "Research Confirms ExxonMobil, Koch-funded Climate Denial Echo Chamber Polluted Mainstream Media," Desmog.com, November 23, 2015.

20. "Science or Spin?: Assessing the Accuracy of Cable News Coverage of Climate Science (2014)," Union of Concerned Scientists website, April 2014.

21. Coral Davenport, "Large Companies Prepared to Pay Price on Carbon," *New York Times,* December 5, 2013; D. Cardwell, "Under Pressure, Utility to Study Emissions," *New York Times,* January 14, 2014; Mindy Lubber, "Fortune 500s Hitting the Hill on Earth Day," *Forbes,* April 21, 2015.

22. For example, see Julie Hirschfield Davis, "Obama Recasts Climate Change as a Peril with Far-Reaching Effects," *New York Times,* May 20, 2015; Zoe Schlanger, "Pentagon Report: U.S. Military Considers Climate Change a 'Threat Muliplier' That Could Exacerbate Terrorism," *Newsweek,* October 14, 2014; Katherine Bagley, "Pope Francis' Eight Major Messages in His Own Words," *InsideClimate News,* June 18, 2015; Marcia McNutt, "The Beyond-Two-Degree Inferno," *Science* 349 (2015): 7; Samantha Page, "These World Leaders Agree: We Need a Price on Carbon," ClimateProgress website, October 20, 2015.

23. Suzanne Goldenberg, "Americans Care Deeply About 'Global Warming'—But Not 'Climate Change,'" *The Guardian,* May 27, 2014; S. Lacey, "Poll: Americans' Understanding of Climate Change Increasing with More Extreme Weather, Warmer Temperatures," ClimateProgress website, February 29, 2012; Lawrence C. Hamilton and Mary D. Stampone, "Blowin' in the Wind: Short-term Weather and Belief in Anthropogenic Climate Change," *Weather, Climate, and Society* 5 (2013): 112–19.

24. Anita Pugliese and Julie Ray, "Fewer Americans, Europeans View Global Warming as a Threat," Gallup.com, April 20, 2011.

25. Seth Borenstein, "Poll: U.S. Belief in Warming Rises with Thermometer," EDGE Media Network website, March 5, 2012.

26. E. Maibach et al., *The Francis Effect: How Pope Francis Changed the Conversation about Global Warming,* November 2015, George Mason University Center for Climate Change

Communication and Yale Program on Climate Change Communication.

27. Leiserowitz et al., *Climate Change in the American Mind.*

28. Lydia Saad, "Americans' Concerns About Global Warming on the Rise," Gallup.com, April 8, 2013; Jeffrey Jones, "In U.S., Concern About Environmental Threats Eases," Gallup.com, March 25, 2015.

29. Clare Foran, "Americans Don't Talk About Global Warming Very Often," *National Journal,* April 20, 2015.

Chapter 9. Grappling with Despair

1. Wood's approach is summarized in Carla A. Wise, "Climate Revolutionary," *High Country News,* May 12, 2008.

Chapter 10. Seeking Solace

1. From *Joseph Campbell and the Power of Myth with Bill Moyers* (1988; released by Mystic Fire Video, 1997).

2. Sherman A. Russell and Max Aguilera-Hellweg, "Talking Plants," *Discover Magazine,* April 1, 2002.

Chapter 11. Looking for Solutions

1. Cally Carlswell, "A Historic Moment for the Clean Air Act," *High Country News,* July 21, 2014.

2. John M. Broder, "Greenhouse Gases Imperil Health, E.P.A. Announces," *New York Times,* December 8, 2009.

3. Center for Climate and Energy Solutions, "Climate Change 101: Federal Action," C2ES website, January 2011.

4. Felicity Barringer, "For New Generation of Power Plants, a New Emission Rule from the E.P.A.," *New York Times,* March 27, 2012.

5. "Obama Administration Finalizes Historic 54.5 mpg Fuel Efficiency Standards," National Highway Transportation Safety Administration press release, August 28, 2012; Center for Climate and Energy Solutions, "Federal Action on Climate Change and Clean Energy," C2ES website, February 2013.

6. "Regulations and Standards: Heavy-Duty," Transportation and Climate, EPA website, September 9, 2015; Peter Baker and Coral Davenport, "Obama Orders New Efficiency for Big Trucks," *New York Times,* February 18, 2014.

7. John Broder, "U.S. Caps Emissions in Drilling for Fuel," *New York Times,* April 19, 2012; "Regulatory Actions," Oil and Natural Gas Air Pollution Standards, EPA website, September 23, 2015.

8. Coral Davenport, "White House Unveils Plans to Cut Methane Emissions," *New York Times,* March 28, 2014; "Regulatory Initiatives," Climate Change, What the EPA Is Doing, EPA website, November 4, 2015.

9. "The Clean Power Plan," EPA website, November 4, 2015.

10. Center for Climate and Energy Solutions, "Q&A: EPA Regulation of Greenhouse Gas Emissions from Existing Power Plants," C2ES website, August 2015.

11. John Broder, "Obama Readying Emissions Limits on Power Plants," *New York Times,* June 19, 2013.

12. Samantha Page, "The Legal Battle Over Obama's Carbon Rule Just Got Bigger," ClimateProgress website, November 4, 2015.

13. Michael B. Gerrard, "The Supreme Court's Action Threatens Vital Climate Policies," *Yale Environment 360,* February 17, 2016.

14. Charles Komanoff, "Next to Nothing for Climate in Obama Plan," Carbon Tax Center, June 2, 2014.

15. Ibid.

16. Center for Climate and Energy Solutions, "Climate Change 101: Federal Action," C2ES website.

17. "U.S. Energy-Related Carbon Dioxide Emissions, 2014," U.S. Energy Information Administration website, November 23, 2015.

18. Baker and Davenport, "Obama Orders."

19. Ibid.

20. Stanley Reed, "Europe Vote Sets Back Carbon Plan," *New York Times,* April 16, 2013

21. Rachel Cleetus, "How to Cut Carbon and Save Money: RGGI Delivers Yet Again," Union of Concerned Scientists blog, April 22, 2015; "Cap and Trade Resurrected? Some States Awaken to Its Economic Benefits," *InsideClimate News,* July 12, 2012; Dale Kasler, "Companies Spend $1 Billion in Latest California Carbon Auction," *Sacramento Bee,* February 25, 2015; "CA Carbon Market 'Well-constructed, Strong,'" Environmental Leader website, January 10, 2014.

22. Scott Nystrom and Patrick Luckow, *The Economic, Climate, Fiscal, Power, and Demographic Impact of a National Fee-and-Dividend Carbon Tax* (Regional Economic Models, Inc., and Synapse Energy Economics, Inc., June 9, 2014).

23. Diane Toomey, "How British Columbia Gained by Putting a Price on Carbon," *Yale Environment 360* interview with Stewart Elgie, April 30, 2015.

24. "Why Alberta's Carbon Tax Matters," Bloomberg.com, November 30, 2015.

25. Christina Maza, "Everyone's Favorite Climate Change Fix," *Christian Science Monitor,* October 29, 2015.

26. Sophie Wenzlau, "Putting a Price on Carbon," Worldwatch blog, December 9, 2015.

27. C. Komanoff, "Carbon Tax Polling Milestone: 2/3 Support if Revenue-Neutral," Carbon Tax Center, April 15, 2015; Peter D. Howe et al., "Geographic Variation in Opinions on Climate Change at State and Local Scales in the USA," *Nature Climate Change* 5 (2015): 595–603.

28. Center for Climate and Energy Solutions, "Renewable and Alternative Energy Portfolio Standards," U.S. Climate Policy Maps, C2ES website, July 29, 2015.

29. Bobby Magill, "Hawaii, Vermont Set Ambitious Examples for Renewables," Climate Central website, July 2, 2015.

30. James Cronin, "Oregon Lawmakers Pass Landmark Bill to Eliminate Coal and Double Renewable Energy," *Portland Business Journal,* March 3, 2016.

31. Maria Gallucci, "Cap and Trade Gives Massachusetts Economy Critical Boost, Defying Naysayers," *InsideClimate News,* December 14, 2011.

32. Lucas Bifera, "Regional Greenhouse Gas Initiative," Center for Climate and Energy Solutions website, December 2013.

33. "Trading Program Linked to Significant Emissions Reductions," Duke University press release, posted on Phys.org, August 24, 2015.

34. Maria Gallucci, "In New Jersey Solar Decision, Economics Trumped Ideology," *InsideClimate News,* August 2, 2012.

35. Joe Cutter, "New Solar Energy Incentives Sought by NJ Lawmaker," New Jersey 101.5 website, August 13, 2013.

36. Roger R. Drouin, "How Conservative Texas Took the Lead in U.S. Wind Power," *Yale Environment 360,* April 9, 2015.

37. Ibid.

38. Stephanie Kirchgaessner, "Mayor Bill de Blasio Pledges to Cut New York Carbon Emissions by 40% by 2030," *The Guardian,* July 21, 2015.

39. "San Francisco Delivers on City Blueprint to Tackle Climate Change," ClickGreen, February 14, 2014.

40. Gennady Sheyner, "Palo Alto Goes 'Carbon Neutral' with Electricity," *Palo Alto Weekly,* March 4, 2013; Erik G. Fowler, Douglas Miller, and Brett Bridgeland, "A California Community Shoots for the Moon: Palo Alto Considers Carbon Neutrality," RMI Outlet, December 3, 2014.

41. Josh O'Gorman, "Vermont Joins Green Coalition," *Rutland Herald,* May 22, 2015.

42. Carla Marinucci, "Good News, Jerry Brown: CA Again World's Eighth Largest Economy," *San Francisco Chronicle,* Politics blog, July 11, 2013.

43. Mark Hertsgaard, "California Takes the Lead with New Climate Initiatives," *Yale Environment 360,* March 8, 2012.

44. Adam Nagourney, "California Governor Orders New Target for Emissions Cuts," *New York Times,* April 29, 2015.

45. Kasler, "Companies Spend $1 Billion."

46. Alan Durning and Yorum Bauman, "17 Things to Know about California's Carbon Cap," Sightline Daily website, May 22, 2014.

47. Hertsgaard, "California Takes the Lead."

48. Silvio Marcacci, "California's Grid Sets Two New Solar Energy Records in Two Days," CleanTechnica website, March 13, 2014.

49. Cheryl Katz, "Surge in Renewables Remakes California's Energy Landscape," *Yale Environment 360,* May 26, 2015.

50. Ibid.

51. Center for Climate and Energy Solutions, "Climate Change 101: Federal Action."

52. Beth Gardiner, "Companies Adopt Green Policies on Their Own," *New York Times,* November 29, 2010; Maria Gallucci, "Major Corporations Quietly Reducing Emissions—and Saving Money," InsideClimate News, September 6, 2012.

53. Coral Davenport, "Industry Awakens to Threat of Climate Change," *New York Times,* January 23, 2014.

54. Ken Silverstein, "Rift Widening Between Energy and Insurance Industries Over Climate Change," *Forbes*, May 18, 2014.

55. Laura Barron-Lopez, "Nike, IKEA join 221 Companies in Backing EPA's Climate Rule," *The Hill*, December 2, 2014.

56. Coral Davenport, "Large Companies Prepared to Pay Price on Carbon," *New York Times*, December 5, 2013.

57. "Putting a Price on Risk: Carbon Pricing in the Corporate World," CDP Report v.1.3, September 2015.

58. Jennifer Rankin, "Europe's Energy Big Six Say Gas Must Help the Fight Against Climate Change," *The Guardian*, June 1, 2015.

59. Maria Gallucci, "Six Major U.S. Banks Urge Global Leaders to Adopt Climate Change Agreement," *International Business Times*, September 28, 2015.

60. Coral Davenport, "Nations Approve Landmark Climate Accord in Paris," *New York Times*, December 12, 2015.

61. Sophie Wenzlau, "The Emissions Gap," Worldwatch blog, December 18, 2015.

62. PricewaterhouseCoopers (PwC), *Low Carbon Economy Index 2015*, October 2015.

63. Davenport et al., "Inside the Paris Climate Deal."

Chapter 12. Spotting Hopeful Trends

1. "Deforestation and Global Warming," Union of Concerned Scientists website, December 9, 2013.

2. Doug Boucher et al., *Deforestation Success Stories* (Union of Concerned Scientists, June 2014).

3. Boucher et al., *Deforestation Success;* Daniel Nepstad et al., "Slowing Amazon Deforestation Through Public Policy and Interventions in Beef and Soy Supply Chains," *Science* 344 (2014): 1118–23.

4. Boucher et al., *Deforestation Success.*

5. P. Friedlingstein et al., "Update on CO2 Emissions," *Nature Geoscience* 3 (2010): 811–12.

6. "About REDD+," UN-REDD Programme website.

7. Boucher et al., *Deforestation Success.*

8. Ibid.

9. Daniel Nepstad et al., "The End of Deforestation in the Brazilian Amazon," *Science* 326 (2009): 1350–51.

10. Associated Press, "Rain Forest Concerns Change Palm Oil Purchasing by Kellogg," *New York Times,* February 18, 2014.

11. Jim Leape, "It's Happening, But Not in Rio," *New York Times,* Opinion, June 24, 2012.

12. Rhett Butler, "A New Leaf in the Rainforest: Longtime Villain Vows Reform," *Yale Environment 360,* March 10, 2014.

13. Associated Press, "Rain Forest Concerns," and S. Smith, "We Did It! Three Companies Go Deforestation-Free," Union of Concerned Scientists email, April 18, 2014.

14. Rhett A. Butler, "Cargill Commits to Zero Deforestation Across Entire Global Supply Chain: All Commodities," Mongabay.com, September 24, 2014.

15. Lisa Palmer, "Will Indonesian Fires Spark Reform of Rogue Forest Sector?" *Yale Environment 360,* November 5, 2015.

16. Rhett A. Butler, "A Conservationist Sees Signs of Hope for the World's Forests," *Yale Environment 360,* October 30, 2014.

17. Boucher et al., Deforestation Success.

18. Ibid.

19. Ibid.

20. Justin Gillis, "Restored Forests Breathe Life into Efforts against Climate Change," *New York Times,* December 23, 2014.

21. Boucher et al., *Deforestation Success.*

22. F. Bongers et al., "The Potential of Secondary Forests," *Science* 348 (2015): 642–43; John Carey, "A Scientist Extols the Value of Forests Shaped by Humans," interview with Susanna Hecht, *Yale Environment 360,* April 5, 2011.

23. Richard Schiffman, "What Lies Behind the Recent Surge of Amazon Deforestation," interview with Philip Fearnside, *Yale Environment 360,* March 9, 2015.

24. David Jolly, "Top U.N. Official Warns of Coal Risks," *New York Times,* November 18, 2013; Christophe McGlade and Paul Ekins, "The Geographical Distribution of Fossil Fuels Unused When Limiting Global Warming to 2 Degrees C," *Nature* 517 (2015):187–90.

25. Kevin Begos, "AP Impact: CO2 Emissions in U.S. Drop to 20-year Low," AP, The Big Story website, August 16, 2012.

26. Ed King, "U.S. Carbon Emissions Set to Fall to Lowest Level in Two Decades," *The Guardian,* April 10, 2015.

27. Mark Hertsgaard, "The Biggest Climate Victory You Never Heard Of," *Al Jazeera English,* Opinion, May 27, 2012.

28. David Roberts, "Yes, Coal Is Dying, but No, EPA Is Not the Main Culprit," *Grist,* October 8, 2012.

29. Ibid.; "U.S. Energy-Related Carbon Dioxide Emissions, 2014," U.S. Energy Information Administration website, November 23, 2015; Hertsgaard, "The Biggest Climate Victory"; Justin Doom, "U.S. Renewable-Energy Capacity Doubled from 2009–2012, BNEF Says," *Bloomberg,* January 31, 2013.

30. Michael Grunwald, "Inside the War on Coal," *Politico,* May 31, 2015.

31. King, "U.S. Carbon Emissions."

32. Bobby Magill, "Flurry of Coal Power Plant Shutdowns Expected by 2016," Climate Central website, February 19, 2014; King, "U.S. Carbon Emissions."

33. Hertsgaard, "The Biggest Climate Victory"; Grunwald, "Inside the War."

34. "An Important Step on Global Warming," *New York Times* editorial, September 22, 2013; Ryan Koronowski, "8 Things You Should Know About the Biggest Thing a President's Ever Done on Climate Change," ClimateProgress, June 2, 2014.

35. "How much U.S. energy consumption and electricity generation comes from renewable sources?" FAQs, U.S. Energy Information Administration website, March 2015.

36. Amory Lovins, "A Bright Future," *Science* 350 (2015): 169.

37. "California State Energy Profile," U.S. Energy Information Association website, October 15, 2015.

38. Lazard LLC, *Lazard's Levelized Cost of Energy Analysis – Version 9.0,* November 2015; Paul Krugman, "Wind, Sun and Fire," *New York Times,* February 1, 2016.

39. "How much of world energy consumption and electricity generation is from renewable sources?" FAQs, U.S. Energy Information Administration website, December 2014.

40. "Nine Surprisingly Strong Signs of Momentum for Climate Action," Union of Concerned Scientists website, May 2015.

41. Oliver Milman, "Zuckerberg, Gates and the Other Tech Titans Form Clean Energy Investment Coalition," *The Guardian,* November 30, 2015.

42. Mark Z. Jacobson et al., "100% Clean and Renewable Wind, Water, and Sunlight (WWS) All-Sector Energy Roadmaps for the 50 United States," *Energy Environ. Sci* 8 (2015): 2093–2117.

43. Alice Friedemann, "Jacobson and Delucchi Energy Dreams Are Irresponsible Fairy Tales," energyskeptic.com, June 2, 2015.

44. Katherine Bagley, "Infographic: A Field Guide to the U.S. Environmental Movement," *InsideClimate News,* April 7, 2014.

45. Grunwald, "Inside the War."

46. Beyond Coal, Sierra Club website.

47. Adam Beitman, "New Analysis by Sierra Club: U.S. on Track to Drive Lowest Emissions Since 1995," Sierra Club press release, November 4, 2015.

48. Lisa Palmer, "Facing Tough Market at Home, U.S. Coal Giant Pushes Overseas," *Yale Environment 360,* July 29, 2013; Tim Dickinson, "How the U.S. Exports Global Warming," *Rolling Stone,* February 3, 2014.

49. Katherine Bagley, "Losing Streak Continues for U.S. Coal Export Terminals," *InsideClimate News,* January 12, 2015.

50. Rob Davis, "Oregon Department of State Lands Rejects Ambre Energy Coal Export Permit, Dealing Major Blow," *Oregonian,* August 18, 2014.

51. Bagley, "Losing Streak."

52. Ibid.

53. "TransCanada Proposes Keystone Oil Pipeline Project," TransCanada press release, February 9, 2005.

54. Bagley, "Infographic: A Field Guide."

55. Coral Davenport, "Citing Climate Change, Obama Rejects Construction of Keystone XL Oil Pipeline," *New York Times,* November 6, 2015; Lisa Hymas, "The 7 Things You Need to Know Now About the Keystone XL Pipeline," Grist, November 6, 2015.

56. Marc Gunther, "Why the Fossil Fuel Divestment Movement May Ultimately Win," *Yale Environment 360,* July 27, 2015.

57. "Divestment Commitments," GoFossil Free.org.

58. John Schwartz, "Norway Will Divest from Coal in Push Against Climate Change," *New York Times,* June 5, 2015.

59. Brett Fleishman, "The California State Legislature Just Passed a Coal Divestment Bill—What Will This Mean for the Global Divestment Movement?" GoFossilFree.org, September 4, 2015.

60. "Divestment Commitments."

61. Y. Bauman and S-L Hsu, "The Most Sensible Tax of All," *New York Times*, Opinion, July 4, 2012; Robert H. Frank, "Carbon Tax Silence, Overtaken by Events," *New York Times*, August 25, 2012; David Kestenbaum, "Economists Have a One-Page Solution to Climate Change," Morning Edition, National Public Radio, June 28, 2013; N. G. Mankiw, "A Carbon Tax That America Could Live With," *New York Times*, August 31, 2013.

62. Chelsea Harvey, "These Could Be the First U.S. States to Tax Carbon—and Give Their Residents a Nice Paycheck," *Washington Post*, November 10, 2015.

63. "Carbon Fee and Dividend Explained," Citizens' Climate Lobby website; "REMI Report," Citizens' Climate Lobby website.

64. Julie Gordon, "Canada PM, Provinces Set Outlines of Carbon Pricing Deal," *Reuters*, March 3, 2016.

65. Timothy Cama, "11 Republicans Vow to Fight Climate Change," *The Hill*, September 17, 2015.

66. Cally Carswell, "Who Really Killed Keystone?" *High Country News*, December 7, 2015.

67. Sophie Wenzlau, "The Emissions Gap," Worldwatch blog, December 18, 2015; Bill McKibben, "Falling Short on Climate in Paris," *New York Times*, December 13, 2015.

68. Eric Roston, "The Caucus to Save the World," *Bloomberg*, February 11, 2016.

69. Dan Vergano, "Attorneys General Will Investigate Energy Firms for Climate Deception," *BuzzFeed*, March 29, 2016.

70. For a full description of the litigation, see Mary Christina Wood, *Nature's Trust: Environmental Law for a New Ecological Age* (Cambridge University Press, 2013), 220–229.

71. Gabriel Nelson, "Young Activists Sue U.S., States Over Greenhouse Gas Emissions," *New York Times*, May 5, 2011.

72. Press Release, "Victory in Landmark Climate Case," Our Children's Trust, April 8, 2016.

73. Press Release, "Youths Secure Second Win in Washington State Climate Lawsuit," Our Children's Trust, April 29, 2016.

Chapter 13. Preparing

1. Jim and Janice Leach, "Home Gardening on the Rise in the U.S.—Gardeners Have a Variety of Motivations," *Ann Arbor News,* January 21, 2012; Amanda Memrick, "Home Gardening on the Rise in Gaston, Nation," *Gaston Gazette* (Gastonia, North Carolina), June 22, 2013.

2. For example, see Fran Korten, "Main Street Comeback: How Independent Stores Are Thriving (Even in the Age of Amazon)," *Yes! Magazine,* October 25, 2013; Megha Satyanarayana, "Urban Farming Invigorates Detroit Neighborhood," *Detroit Free Press,* May 20, 2013.

3. Sarah A. Low et al., "Trends in U.S. Local and Regional Food Systems: A Report to Congress," USDA Economic Research Service, January 2015.

4. Jon Mooallem, "The End Is Near! (Yay!)," *New York Times Magazine,* April 16, 2009.

5. Craig K. Comstock, "The 'Transition Town' Movement's Initial Genius," *Huffington Post,* May 25, 2011.

6. The website has not updated this number since 2013, but the group continues to hold trainings and sponsor activities around the world.

7. See, for example, this story about the Dark Mountain people: Daniel Smith, "It's the End of the World as We Know It … and He Feels Fine," *New York Times Magazine,* April 17, 2014.

8. Wendell Berry, *Our Only World: Ten Essays* (Counterpoint Press, 2015), p. 168.

Chapter 14. Finding My Way

1. James Hansen, "Global Warming: Is There Still Time to Avoid Disastrous Human-Made Climate Change? i.e. Have We Passed a 'Tipping Point'?" Discussion on April 23, 2006, at the National Academy of Sciences, Washington, DC.

2. E. Kintisch, "Hansen's Retirement from NASA Spurs Look at His Legacy," *Science* 340 (2013): 540–41.

3. Naomi Oreskes, "Exxon's Climate Concealment," *New York Times,* Opinion, October 9, 2015; Jody Warrick, "Why Are So Many Americans Skeptical about Climate Change? A

Study Offers a Surprising Answer," *Washington Post,* November 23, 2015; Justin Gillis and John Schwartz, "Deeper Ties to Corporate Cash for Doubtful Climate Researcher," *New York Times,* February 21, 2015.

4. "People's Climate March Draws 400,000," *Science* 345 (2014): 1544.

5. Samantha Page, "These House Representatives Just Did Something Amazing on Climate Change," ClimateProgress website, February 5, 2016.

6. Bill McKibben, "Global Warming's Terrifying New Math," *Rolling Stone,* August 2, 2012.

7. These titles include: Dale Jamieson, *Reason in a Dark Time: Why the Struggle Against Climate Change Failed – And What It Means for Our Future* (Oxford University Press, 2014); George Marshall, Don't Even Think About It: Why Our Brains Are Wired to Ignore Climate Change (Bloomsbury, 2014); Kari Norgaard, *Living in Denial: Climate Change, Emotions, and Everyday Life* (MIT Press, 2011); and Per Espen Stoknes, *What We Think About When We Try Not To Think About Global Warming: Toward a New Psychology of Climate Action* (Chelsea Green, 2015).

8. Berry, *Our Only World: Ten Essays,* 167.

Index

A

Acidification, ocean, xxiv, 21–22, 23

Act, failure to, 66, 67–68, 82

  and resistance to change, 130

Action, climate. *See* Climate action

Activism, climate. *See* Climate activism

Agriculture, xxiii, 43–44. *See also* Crop failures; Food production

  climate-friendly, 47–48

  contributions to climate change, 46–48

  emissions from, 46–47

  genetically modified (GM) crops, 48

  local food systems, 49

  threats to, 42–45

  and tropical deforestation, 108

Amazon rainforest, 34, 110. *See also* Deforestation

Andrews Experimental Forest, 85–93, 132–133

Animal consciousness, 89, 90

Animal farming, 48

Animals, land

  benefiting from climate change, 15–16

  changes in seasonal activities, 12–14

  distribution of, 13–14, 16

  food sources, 18

  multifaceted selection event, 17

Antarctic, 26

Apathy, 82

Arctic
- decline of bird species in, 17
- effects of warming in, xxiii
- ice in, 18, 26, 27 (figure)
- wildfires in, 37

Atmospheric Trust Litigation, 118
Automobiles, 73, 96–97

B

Bangladesh, 28
Beauty, natural, 80–81
Beef industry, 108. *See also* Meat consumption
Berry, Wendell, 85–86, 89, 125, 138
Beyond Coal campaign, 110–111, 113
Bezos, Jeff, 112
BioScience, 54
Birds, 12–13, 14, 15–16, 18, 26
Blame, for climate change, 63–68
Bleaching, coral, 23
Bloomberg, Michael, 113
Boundaries, 88–89
Brazil, 107–108, 109, 110
Breakthrough Energy Coalition, 112
Burns, controlled, 36. *See also* Forest fires
Bush, George H. W., xiii
Bush, George W., xxx (box)
Business. *See* Corporations
Butterflies, 13, 14

C

Caldeira, Ken, 24
California
cap-and-trade program, 99, 103
- Global Warming Solutions Act, 102–103
- renewable energy in, 111

Campbell, Joseph, 88
Canada, 117
Capacity building, 133–134
Cap-and-trade programs, 98–99, 101, 103
Carbon dioxide, xx, xxi
- and food production, 42–43
- measuring, 32
- oceans' absorption of, 21
- releases of, 34
- trees' absorption of, 32–33

Carbon Pricing Leadership Coalition, 100
Carbon-pricing schemes, 98–100
- cap-and-trade programs, 98–99, 101, 103
- carbon taxes, 98–100, 104, 116, 117, 117 (box), 135

Carbon Tax Center, 100

Carbon taxes, 98–100, 104, 116, 117, 117 (box), 135
Care, failure to, 82
CDP, 104
Change
  opportunity for, 57–58
  possibility of, 107
  resistance to, 130
China, 111
Citizens' Climate Lobby (CCL), 100, 113, 116, 117 (box), 135–137, 138
Clean Air Act, 96–97
Clean Power Plan, 68, 97, 104, 111
Climate action. *See also* Act, failure to; Carbon-pricing schemes
  barriers to, 67–68
  carbon-pricing schemes, 98–100
  Clean Power Plan, 68, 97, 104, 111
  by corporations, 103–104
  and Keystone XL pipeline, 68, 114–115, 118
  Kyoto Protocol, xiii
  need for, 68
  and need for engagement, 134–135
  Paris climate accord, xiv, xxvii, 69, 118
  Paris Climate Conference, 100, 104
  portrayal of, 66
  and slowing of deforestation, 108–109
  state/local actions, 100–103
  by U.S. government, 96, 118, 129
Climate activism
  Beyond Coal campaign, 110–111, 113
  carbon taxes campaign, 116, 117 (box)
  Citizens' Climate Lobby, 100, 113, 116, 117 (box), 135–137, 138
  fossil fuels divestment movement, 115–116
  halting coal exports, 113–114
  and Keystone XL pipeline, 68, 114–115, 118
  litigation, 118–119
  organizations, 141
  rise of, 112–117
  350.org, 113, 114, 115, 135
Climate attribution science, 9–10
Climate change
  Americans' beliefs/attitudes about, 69–70, 70 (figure)
  coming to terms with, 30
  defined, xiv (box)
  effects of, xxii–xxiii, xxviii, 60
  and human activity, xxii
  increased acceptance of, 129
  views on causes of, 69

Climate change deniers, xxii, 56, 64–65
  on Antarctic sea ice, 26
  in media, 65–66
  as minority, 68
Climate Cover-Up (Hoggan and Littlemore), 64
Climate fatigue, 66
Climate scientists, 55, 65
Climate sensitivity, xxvi
Climate Solutions Caucus, 118
Clinton, Bill, xiii
Coal
  Beyond Coal campaign, 110–111, 113
  decline in coal burning, 110–111
  halting exports, 113–114
Cold events, 6, 7
Companies. *See* Corporations
Congo Basin, 109
Congress. *See also* Government, U.S.; Politicians
  climate action by, 118, 129
  efforts to intimidate climate scientists, 65
Consensus, scientific
  acceptance of, 69
  on climate change, xx–xxiv, 55, 63
  forming, xix–xx
Consumption, 75
Conway, Erik, 64
Corals, 23–24
Corporations
  climate action by, 103–104
  misinformation spread by, 129
  and slowing of deforestation, 108–109
Corvallis, Ore., 52, 124–125
Corvallis Sustainability Coalition, 124–125
Crop failures, 7, 29, 42, 43. *See also* Agriculture; Food production
Cyclones, 5–6, 8, 28. *See also* Weather, extreme
Cynicism, 66, 68

D

Dairy consumption, 47, 48
Death
  fear of, xvi
  and weather, 4–6, 7, 9
The Death of Nature (McKibben), 88
Deforestation, 33, 107–110. *See also* Forests; Trees
Despair, 79–83, 133
Disbelief, in climate change science, 55
Divoky, George, 12–13, 18
Driving, 73, 74
Droughts, xxiii, 5, 7. *See also* Weather, extreme
  and forest death, 34
  threat to agriculture, 44

E

Ehrlich, Paul, 53
Emissions. *See* Greenhouse gases
Endangerment finding, 96–97
Energy, 110. *See also* Coal; Renewable energy
Energy Information Administration, U.S., 111
Engagement, 134–135. *See also* Climate activism
Environmentalist's paradox, 53–55
Environmental Protection Agency (EPA), 96, 104, 110, 111
Evolution, 132
Extinctions, xxiii, 16–17, 128

F

Farming, 41. *See also* Agriculture; Food production
Farmland
  abandoned, 38
  available, 44
Fertilizer, 47
Fires, 4–5, 35–38, 80–81. *See also* Weather, extreme
Flooding, xxiii, 5, 28. *See also* Weather, extreme
Flying foxes, 14–15
Food and Agriculture Organization (FAO), 46–47
Food insecurity, 45
Food prices, 42, 45
Food production, xxiii, 42–44. *See also* Agriculture; Crop failures; Hunger
Food systems, local, 49. *See also* Relocalization efforts
Forest fires, 4–5, 35–38, 80–81. *See also* Weather, extreme
Forest management, 36
Forests, xxiii. *See also* Trees
  Amazon rainforest, 34
  and carbon dioxide absorption, 33
  deforestation, 33, 107–110
  die offs, 15, 34–35, 38–39
  REDD+ program, 108
  regrowth, 38, 108, 109–110
  thinning of, 36–37
  threats to, 34
  types of, 33
Fossil fuel industry, 70–71, 129
Fossil fuels, xxix, 64. *See also* Coal
Fossil fuels divestment movement, 115–116
Foxes, flying, 14–15
Fracking, 110
Francis (pope), 69
Future, 124, 138–139
  preparing for, 121–126

G

Gates, Bill, 48, 112
Genetically modified (GM) crops, 48
Gilding, Paul, 55
Gillis, Justin, 8
Glaciers, xxiii
Global mean land-ocean temperature index, xx–xxi, xxi (figure)
Global surface temperatures, xxv, xxv (figure)
Global warming. *See also* Climate change
  defined, xiv (box)
  observed, xx
Goddard Institute for Space Studies, xx, xxx (box)
Government, U.S. *See also* Congress; Politicians
  climate action by, 96, 118, 129
  and corporate money, 82
  fossil fuel industry's control of, 70–71
  opposition to climate action, 97, 98, 99
Governments, and slowing of deforestation, 109
Grassroots campaigns. *See* Climate activism
Greenhouse effect, xx
Greenhouse gases, xx, xxvi
  and deforestation, 107
  effects of, xxii
  federal policies on, 97–100
  and human activities, xxi
  lag time in effects of, xxiii
  limiting, xxviii
  and limiting increase in warming, 104–105
  and livestock, 46–47
  measuring, xxi
  reducing pace of, 98
  reduction targets, 102–103
  regulation of, 96
  state/local action on, 100–103
  and tar sands mining, 114
Grief, 83. *See also* Despair
Grist, 59
Guillemots, 12–13, 18, 26
Guilt, 74–77
Gulf Stream current, 29

H

Hansen, James, xxix, xxx (box), 115, 127–128, 130
Heartbreak, 83. *See also* Despair
Heat waves, xxiii, 5, 6, 9, 14–15. *See also* Weather, extreme
Helplessness, 91
Hoggan, James, 64
Homesteading skills, 121–122, 126

Hope, 107, 127
Hopkins, Rob, 122–123
Hot events, 6. *See also* Heat waves; Weather, extreme
Human Development Index, 54
Hunger, 44, 45
Hurricanes, 8

I

Ice sheets, xxiii, 16, 25–27
India, 109
Individual, helplessness as, 91
Indonesia, 107, 109
Industry. *See* Corporations
Intergovernmental Panel on Climate Change (IPCC), xxiv (box), xxv, xxvii, 55, 63
    on agriculture, 42, 43, 44
    on extreme weather, 4
    on sea level rise, 28
International Energy Agency (IEA), xxix
IPCC (Intergovernmental Panel on Climate Change). *See* Intergovernmental Panel on Climate Change

K

Keeling, Charles David, 32
Keeling Curve, 32
Kenward, Alyson, 10
Keystone XL pipeline, 68, 114–115, 118
Klose, Stefan, 15
Koch brothers, 65
Kyoto Protocol, xiii

L

Litigation, 118–119
Littlemore, Richard, 64
Livestock, and greenhouse gas emissions, 46–47
Logging, 36–37
Low Carbon Economy Index, 104

M

Madagascar, 109
Malaysia, 109
Marmots, 14
Marshall, George, 67
McKibben, Bill, 49, 64, 88, 95, 115, 130, 131
Meat consumption, 47, 48
Media, 65–67, 70, 95
Merchants of Doubt (Oreskes and Conway), 64
Monsanto, 48
Motherhood, xvi, 81–82

N

NASA, 25, 44. *See also* Goddard Institute for Space Studies
National Climate Assessment, Third, 28

National Oceanic and Atmospheric Administration (NOAA), 4, 5
National Research Council, 37
Natural gas production, 110
Natural selection, 132
Natural world, 87
Nature, 9
Nature Geoscience, xxviii
NCAR, 10
New Jersey, 101–102
News, bad, 66
New York City, 51–52, 102
New York Times, 8, 31
Nitrous oxide emissions, 47
Norgaard, Kari, 67

O

Obama, Barack, 63, 115, 117, 133
Ocean acidification, xxiv, 21–22, 23
Ocean currents, 20–21, 27, 29
Oceans
  absorption of carbon dioxide, 21
  coral, 23–24
  food webs, 23
  importance of, 20
  marine organisms, 22–24
  and melting ice, 25–27
  sea level rise, xxiii, xxiv, 22, 25, 27–29
  sea surface temperatures, 8, 21, 22
Oil
  Keystone XL pipeline, 68, 114–115, 118
Opportunity, for change, 57–58
Optimism, 138
Oregon, 124–125
Oreskes, Naomi, 64
Organisms, marine, 22–24
Organizations, 141
Our Children's Trust, 118

P

Palm oil industry, 108, 109
Palo Alto, Calif., 102
Paradox
  defined, 53
  environmentalist's paradox, 53–55
  on future, 138
  of scientific certainty and public disbelief, 55
  of twin possibilities, 56–58
Paris climate accord, xiv, xxvii, 69, 118
Paris Climate Conference, 100, 104
"The Peace of Wild Things" (Berry), 85–86, 89
Perspective, 87
Pests, agricultural, 44

Phytoplankton, 23. *See also* Oceans
Pines, whitebark, 15
Plant consciousness, 89–90
Plants
communication between, 89–90
distribution of, 13–14
Polar bears, 14, 18
Politicians, 129. *See also* Congress; Government, U.S.
efforts to intimidate climate scientists, 65
opposition to climate action, 97, 99
response to climate change, xiii
Poverty, 45
Power plants, 97
Preparation, 121–126, 133–134
Purchases, 75

R

Rain, 7. *See also* Weather, extreme
Reducing Emissions from Deforestation and Forest Degradation in Developing Countries (REDD+), 108
Reforestation, 38, 108, 109–110
Regional Greenhouse Gas Initiative (RGGI), 99, 101
Reilly, William, xiii
Relocalization efforts, 121, 122, 124, 126
Renewable energy, 101, 111–112. *See also* Solar power; Wind power
Resilience, building, 122–123
Resources, consumption of, 53
Responses, to climate change
blame, 63–68
despair, 79–83, 133
guilt, 74–77
helplessness, 91
and likelihood of failure, 131
looking for solutions, 95
possibilities, xiv
seeking solace, 85–93
self-sufficiency, 121–122
technology-driven, 48
Reynolds, Mark, 138
RGGI (Regional Greenhouse Gas Initiative), 99, 101
Robbins, Jim, 31, 35
Roberts, David, 59

S

Sandy (hurricane), 28
San Francisco, 102
Scale, importance of, 90–92
Science, 34, 66
Sea ice, melting of, 16, 25–27, 27 (figure)

Sea level rise, xxiii, xxiv, 22, 25, 27–29. *See also* Oceans

Sea surface temperatures, 8, 21, 22

Self-sufficiency, 121–122

Sierra Club, 110–111, 113

Snowfall, 5, 7

Solace, seeking, 85–93

Solar power, 101–102, 111

Solutions. *See also* Climate action; Climate activism

- looking for, 95
- technology-driven, 48

Soon, Wei-Hock, 66

Soy industry, 108

Standard of living, 53

Starvation, 45

Stoknes, Per Espen, 67

Sumatra, 109

Sustainability, 124

Swanson, Fred, 132–134

T

Take Charge Corvallis, 125

Tar sands mining, 114

TEDx talk, 59

Temperatures

- average, xxi
- and crop failures, 42
- global mean land-ocean temperature index, xx–xxi, xxi (figure)
- global surface temperatures, xxv, xxv (figure)
- limiting increase in, xxvii–xxviii, 104–105
- sea surface temperatures, 8, 21, 22

Texas, 102, 111

Third National Climate Assessment, 28

350.org, 113, 114, 115, 135

Tipping point, climate, 127–128

Tornadoes, 8

Transition Movement, 122–123, 124

Travel, 74, 76–77

Trees. *See also* Forests

- absorption of carbon dioxide, 32–33
- forest fires, 4–5, 35–38, 80–81
- importance of, 31
- mortality rates, 110
- planting, 38
- threats to, 34

Trenberth, Kenneth, 10

Typhoons, 8. *See also* Cyclones

U

Uncertainties, xxv–xxvii, xxv (figure), xxviii (box)

- about cyclones, 8
- about tornadoes, 8
- and agriculture/food production, 44
- and forest die offs, 38–39
- and sea level rise, 28

Union of Concerned Scientists, 107

United Nations, 108
United States. *See also* Government, U.S.
  cap-and-trade programs in, 101, 103
  climate action by corporations in, 103–104
  decline in coal burning in, 110–111
  state/local climate action in, 100–103

W

Wagner, Eric, 53–54
Warming. *See also* Temperatures
  future, xxvii–xxviii
Warnings, xxx (box), 127–128
Water, 20. *See also* Oceans
Weather, extreme, xxiv, 4, 129
  and acceptance of climate change, 69
  and combination of natural variability and climate change, 10
  and crop failures, 42, 43
  cyclones, 5–6, 8, 28
  and deaths, 4–6, 7, 9
  fires, 4–5, 35–38, 80–81
  increase in, 6, 128
  link to climate change, 6–8, 9–10
Welbergen, Justin, 15
Whitebark pines, 15
Whitty, Julia, 20, 30
Wildfires, 4–5, 35–38, 80–81. *See also* Weather, extreme
Willamette Valley, 124
Wind power, 97–98, 102
Wood, Mary Christina, 79

Z

Zuckerberg, Mark, 112

About the Author

Carla A. Wise is an environmental writer, climate activist, and mom who has worked as a teacher, natural resource policy analyst, environmental consultant, and plant conservation biologist. She holds a master's degree in environmental studies and a Ph.D. in plant conservation biology. Her writing has appeared in the *Oregonian, High Country News,* the *Huffington Post,* the *Utne Reader,* and elsewhere. She loves hiking, floating rivers, traveling, gardening, and hanging out with her husband, daughter, and two dogs at their home in Corvallis, Oregon. Visit her at www.wiseonearth.com.